COMPUTER METHODS FOR THE RANGE OF FUNCTIONS

ELLIS HORWOOD SERIES IN MATHEMATICS AND ITS APPLICATIONS
Series Editor: Professor G. M. BELL, Chelsea College, University of London

Statistics and Operational Research
Editor: B. W. CONOLLY, Chelsea College, University of London

Baldock, G. R. & Bridgeman, T.	Mathematical Theory of Wave Motion
de Barra, G.	Measure Theory and Integration
Berry, J. S., Burghes, D. N., Huntley, I. D., James, D. J. G. & Moscardini, A. O.	Teaching and Applying Mathematical Modelling
Burghes, D. N. & Borrie, M.	Modelling with Differential Equations
Burghes, D. N. & Downs, A. M.	Modern Introduction to Classical Mechanics and Control
Burghes, D. N. & Graham, A.	Introduction to Control Theory, including Optimal Control
Burghes, D. N., Huntley, I. & McDonald, J.	Applying Mathematics
Burghes, D. N. & Wood, A. D.	Mathematical Models in the Social, Management and Life Sciences
Butkovskiy, A. G.	Green's Functions and Transfer Functions Handbook
Butkovskiy, A. G.	Structure of Distributed Systems
Chorlton, F.	Textbook of Dynamics, 2nd Edition
Chorlton, F.	Vector and Tensor Methods
Dunning-Davies, J.	Mathematical Methods for Mathematicians, Physical Scientists and Engineers
Eason, G., Coles, C. W., Gettinby, G.	Mathematics and Statistics for the Bio-sciences
Exton, H.	Handbook of Hypergeometric Integrals
Exton, H.	Multiple Hypergeometric Functions and Applications
Exton, H.	q-Hypergeometric Functions
Faux, I. D. & Pratt, M. J.	Computational Geometry for Design and Manufacture
Firby, P. A. & Gardiner, C. F.	Surface Topology
Gardiner, C. F.	Modern Algebra
Gasson:	Geometry of Spatial Forms
Goodbody, A. M.	Cartesian Tensors
Goult, R. J.	Applied Linear Algebra
Graham, A.	Kronecker Products and Matrix Calculus: with Applications
Graham, A.	Matrix Theory and Applications for Engineers and Mathematicians
Griffel, D. H.	Applied Functional Analysis
Hoskins, R. F.	Generalised Functions
Hunter, S. C.	Mechanics of Continuous Media, 2nd (Revised) Edition
Huntley, I. & Johnson, R. M.	Linear and Nonlinear Differential Equations
Jaswon, M. A. & Rose, M. A.	Crystal Symmetry: The Theory of Colour Crystallography
Kim, K. H. and Roush, F. W.	Applied Abstract Algebra
Kosinski, W.	Field Singularities and Wave Analysis in Continuum Mechanics
Marichev, O. I.	Integral Transforms of Higher Transcendental Functions Using Computers
Meek, B. L. & Fairthorne, S.	Using Computers
Muller-Pfeiffer, E.	Spectral Theory of Ordinary Differential Operators
Nonweiler, T. R. F.	Computational Mathematics: An Introduction to Numerical Analysis
Oldknow, A. & Smith, D.	Learning Mathematics with Micros
Ogden, R. W.	Non-linear Elastic Deformations
Ratschek, H. & Rokne, Jon	Computer Methods for the Range of Functions
Scorer, R. S.	Environmental Aerodynamics
Smith, D. K.	Network Optimisation Practice: A Computational Guide
Srivastava, H. M. and Manocha, H. L.	A Treatise on Generating Functions
Temperley, H. N. V. & Trevena, D. H.	Liquids and Their Properties
Temperley, H. N. V.	Graph Theory and Applications
Thom, R.	Mathematical Models of Morphogenesis
Thomas, L. C.	Games Theory and Applications
Townend, M. Stewart	Mathematics in Sport
Twizell, E. H.	Computational Methods for Partial Differential Equations
Wheeler, R. F.	Rethinking Mathematical Concepts
Willmore, T. J.	Total Curvature in Riemannian Geometry
Willmore, T. J. & Hitchin, N.	Global Riemannian Geometry

COMPUTER METHODS FOR THE RANGE OF FUNCTIONS

H. RATSCHEK
Professor of Mathematics
Mathematisches Institute der Universität
Düsseldorf, West Germany

and

J. ROKNE, B.Eng., M.Sc., Ph.D.
Professor of Computer Science
University of Calgary, Calgary
Alberta, Canada

ELLIS HORWOOD LIMITED
Publishers · Chichester

Halsted Press: a division of
JOHN WILEY & SONS
New York · Chichester · Brisbane · Toronto

First published in 1984 by
ELLIS HORWOOD LIMITED
Market Cross House, Cooper Street, Chichester, West Sussex, PO19 1EB, England

The publisher's colophon is reproduced from James Gillison's drawing of the ancient Market Cross, Chichester.

Distributors:

Australia, New Zealand, South-east Asia:
Jacaranda-Wiley Ltd., Jacaranda Press,
JOHN WILEY & SONS INC.,
G.P.O. Box 859, Brisbane, Queensland 40001, Australia

Canada:
JOHN WILEY & SONS CANADA LIMITED
22 Worcester Road, Rexdale, Ontario, Canada.

Europe, Africa:
JOHN WILEY & SONS LIMITED
Baffins Lane, Chichester, West Sussex, England.

North and South America and the rest of the world:
Halsted Press: a division of
JOHN WILEY & SONS
605 Third Avenue, New York, N.Y. 10016, U.S.A.

©1984 H. Ratschek and J. Rokne/Ellis Horwood Limited.

British Library Cataloguing in Publication Data
Ratschek, H.
Computer methods for the range of functions. –
(Ellis Horwood series in mathematics and its applications)
1. Functions 2. Approximation theory
I. Title II. Rokne, J.
515.7   QA331

Library of Congress Card No. 84-3737

ISBN 0-85312-703-4 (Ellis Horwood Limited)
ISBN 0-470-20034-0 (Halsted Press)

Printed in Great Britain by Butler & Tanner, Frome, Somerset..

COPYRIGHT NOTICE –
All Rights Reserved. No part of this publication may be reproduced, stored in a retrieval system, or transmitted, in any form or by any means, electronic, mechanical, photocopying, recording or otherwise, without the permission of Ellis Horwood Limited, Market Cross House, Cooper Street, Chichester, West Sussex, England.

Table of contents

Preface		7
Chapter I	Interval Arithmetic	12
	1.1 The trick of interval arithmetic	12
	1.2 Interval arithmetic operations	13
	1.3 Norms and metrics in interval spaces	17
	1.4 Inclusion of the range — a combinatorial problem?	20
	1.5 Inclusion isotonicity and Lipschitz conditions	23
Chapter II	The Standard Centred Form	30
	2.1 Introduction	30
	2.2 The standard centred form for rational functions in one variable	32
	2.3 Symmetric intervals — or not?	37
	2.4 Standard centred forms of higher order for functions in one variable	42
	2.5 Standard centred forms for functions in several variables	49
	2.6 Krawczyk's centred forms	59
Chapter III	General Definition of Centred Forms	63
	3.1 The reasons for a general definition	63
	3.2 General definition of centred forms	64
	3.3 The quadratic convergence	66
	3.4 The standard and Krawczyk's centred forms	72
	3.5 Mean-value and Taylor-forms	75
	3.6 Discussion of the general definition	81
	3.7 Practical considerations for constructing centred forms	85
	3.8 The knowledge about monotonicity	89

Table of Contents

Chapter IV	More About Quadratic Convergence		93
	4.1	The subdivision method	93
	4.2	Global convergence in the subdivision method	95
	4.3	The method of Skelboe	99
Chapter V	Optimality of the Standard Centred Forms		103
	5.1	Computable approximations	103
	5.2	Approximations for the range	105
	5.3	A special case	107
	5.4	The general case	109
Chapter VI	Other Inclusions for the Range of a Function		111
	6.1	Quotients of centred forms	111
	6.2	The circular complex centred form	116
	6.3	The methods of Cargo–Shisha and Rivlin	130
	6.4	The method of Dussel–Schmitt for polynomials	137
	6.5	The method of Hansen	141
	6.6	Centred forms and interval operators	145
	6.7	Remainder and interpolation forms	148
Appendix			157
Bibliography			159
List of Symbols			165
Index of Names			166
Index of Subjects			168

Preface

Information about the range of a function f or related functions (derivative, partial derivative, inverse, etc.) is of great interest to people working in the fields of numerical and functional analysis, differential equations, linear algebra, approximation and optimization theory and other disciplines.

The problem of finding the global minimum (infimum) or the global maximum (supremum) of f over a domain X is, for example, a basic problem of global optimization theory.

The bounds of the range of the derivative are furthermore used to construct Lipschitz constants or other constants which are required in existence or convergence assertions in fixed point theory and iterative procedures.

In order to choose an appropriate algorithm for finding an approximate property of f it is often necessary to have information about the geometry and other properties of f obtained from knowledge of the range of f.

If $\tilde{f}(X)$ denotes the range of f over the domain X, for example, then the set $\tilde{f}(X) \cup X$ contains all fixed points of f and if $\tilde{f}(X) \cup X$ is empty then f has no fixed points, etc. Or, if $\tilde{f}'(X)$ denotes the range of the derivative over X, then $0 \notin \tilde{f}'(X)$ says that f is monotone over X, or if $0 \leq \inf_{x \in X} f''(x)$, then f is convex over X, or if $0 \notin \tilde{f}(X)$ then a programmer can use procedures where it is possible to divide through values $f(x)$ for any $x \in X$, or if the set of values $[f(x) - f(y)]/(x - y)$, $x \neq y$ has an upper bound then this bound is a Lipschitz constant for f, etc.

The aim of this book is to present formulas and methods that may be applied to the problem of finding the range of a function in one or several variables over an interval yielding the range or intervals including the range. This is equivalent to saying that the formulas or methods yield the global minimum (or maximum) of f or lower (upper) bounds for the global minimum (maximum). The reason for choosing outer (and not inner) estimations for the range is that the logic of methods for solving for zeros of equations, etc. require outer estimations.

As an example let $F(X)$ be an outer and $\psi(X)$ an inner estimation of the range $\tilde{f}(X)$, that is,

$$\psi(X) \subseteq \tilde{f}(X) \subseteq F(X).$$

Then, returning to the previous examples, the conclusion

$$\text{if } F(X) \cap X = \emptyset \quad \text{then} \quad \tilde{f}(X) \cap X = \emptyset$$

is valid. The statement '$F(X) \cap X$ contains all fixed points of F' is also valid. If $F(X)$ is replaced by $\psi(X)$, then neither the conclusion nor the statement are valid in general. Similar considerations hold for the other examples that were mentioned.

In this book a theory is treated whose application will guarantee *safe bounds* for the range (or the global minimum, etc.) for *each function considered*. Furthermore, good convergence properties of the bounds to the range are obtained via an iterative approach. The theory is essentially based on two very simple but, nevertheless, extremely effective principles due to Moore (1966).

The *first principle* is the repeated utilization of the fact that for each continuous function $f: X \to R$, $X \in I$ (R denotes the set of reals, I the set of compact intervals of R) an interval $F(X) \in I$ exists with

$$\tilde{f}(X) \subseteq F(X).$$

The important practical point is that such inclusions $F(X)$ not only do exist, but that they can also be obtained both in theoretical investigations and in numerical calculations on a computer. These inclusions are obtained via the use of *interval arithmetic*. Additionally, if interval arithmetic is implemented on a computer then for any function f that can be evaluated by a computer the inclusion $F(X)$ is obtained almost automatically by replacing the variable x of the function procedure by the domain X.

At this point it is necessary to emphasize that the theory and the contents of this book do *not depend* on a computer or on the interval arithmetic equipment of a computer or on any other practical, numerical, or programming aid. However, *the theoretical or historical background of some part of the theory developed in this book needs some theoretical facts of interval arithmetic*.

The additional advantage of the described theory is that the related methods can easily be translated into computer programs.

The programming of the methods is not done in the book, it is left as an option for the reader. The reader is, however, aided in this task by some hints and remarks.

The *second* of the two above-mentioned principles consists of an appropriate technique for constructing inclusions $F(X)$ such that, together with a subdivision method of X, quadratic convergence of the inclusions to the range can be attained by an iterative procedure. This technique is called the centred form. Using this technique an inclusion $F(X)$ is obtained practically in most cases by a Taylor-development of f at the 'centre' of X and replacing the variable x of f by the domain X.

As the manuscript was being completed a totally new method for obtaining inclusions $F(X)$ was brought out by Cornelius–Lohner (1983). The essential feature of this technique is to represent f in the form

$$f = g + r$$

such that g is a 'good' approximation of f and such that the range $\bar{g}(X)$ can be computed exactly (neglecting the rounding errors). If the interval $S(X)$ is an inclusion of $\bar{r}(X)$ then the interval

$$F(X) = \bar{g}(X) + S(X)$$

is an inclusion of $\bar{f}(X)$. This is called a *remainder form*. If g is chosen as a Hermite interpolation polynomial, then $F(X)$ is called an *interpolation form*. The advantage of using interpolation forms is that theoretically any order of convergence can be gained. Convergence order 3 or 4 is recommended in practice.

We have not included methods in our book which are known from global optimization theory since the interested reader can find very good guides to these methods in, for example, McCormick (1972), Spang (1962), Dixon–Gomulka–Szegö (1975), Wilde (1978) and Dixon–Szegö (1978b), and several collections of articles of the subject, for example, Lootsma (1972) and Dixon–Szegö (1975), (1978a). These methods contain grid techniques, random methods, gradient methods, techniques for determining the local minima first and then proceeding to the global minimum, etc. Seemingly none of these methods 'can be guaranteed to locate the global minimum but considerable success has been achieved. It is probably for this reason that random methods for global optimisation are still so popular' (Dixon–Gomulka–Szegö (1975)). *Interval methods, however, guarantee always the location of the global extremum and, in most cases, with arbitrary accuracy.*

Methods for determining the range of functions which are given only implicitly have also not been included in our book. Examples of such functions are given by functions defined by differential equations, integral equations, or other operator equations. Methods for obtaining the range of such functions depend very deeply on the special theories like the theory of differential equations, etc.

We now give a short description of the content of the book. A somewhat better overview would be obtained by reading the introductions at the beginning of most of the chapters and sections.

Chapter I offers an introduction to interval arithmetic and the background for its application for obtaining inclusions of the range. Special attention is given to Lipschitz conditions of interval functions, since these properties are responsible for the quadratic convergence assertions discussed in the sequel.

Chapter II. The standard centred forms are explicit formulas for inclusions of the range of arbitrary rational functions in one or several variables. There are forms of different order. The higher the order the better is the inclusion. Centred forms that are recursively defined via interval slopes are further discussed. These are called Krawczyk's centred forms. They have a lower computational complexity than the standard centred forms.

Chapter III. The standard and Krawczyk's centred forms are applicable for rational functions only. A more general definition of a centred form extends its applicability to non-rational functions as well. However, this generalization gives no explicit formula for concrete inclusions as was the case with the standard centred form. Rather, a criterion is given to show whether or not any given formula can act as a centred form. The main theorem on quadratic convergence is developed and proven with all details. It is shown that Krawczyk's centred form, the standard centred form, the mean-value and Taylor-forms satisfy the general definition of a centred form. An extensive explanation of this general definition follows since there are several differences between this definition and those of other authors. These differences arise mainly with the steps from a centred form to a centred form function and from the 'centre' of the domain to a developing point function. These steps are accomplished in this chapter and are necessary for a clear exposition. Finally, many hints for the application of the centred form are given. For example, the computational complexity of the formulas can be considerably reduced if f is known to be monotone in some of the variables.

Chapter IV. The subdivision method is a procedure to improve the inclusions of the range obtained by some method by subdividing the domain. If the diameters of the subdomains which are generated tend to zero then, under certain conditions on the function f, the inclusions of the range tend to the range quadratically.

Chapter V. It is shown that the standard centred form is an optimal formula for constructing the desired inclusions in the sense that no other formula which processes the same data (information) of f as the standard centred form, is better than the standard centred form.

Chapter VI. Some other methods for constructing inclusions for the range are discussed. They also depend merely on interval ideas or are extensions of the real case to the complex case. Advantages and disadvantages of these methods are discussed. For example, the use of Bernstein polynomials brings out criteria that allows one to decide whether an inclusion of the range is already the range itself. The method of Cornelius–Lohner (1983), that is the construction of interpolation forms, which leads to higher than quadratic convergence, is extensively discussed and explicit formulas for inclusions are given that have a third order convergence.

Acknowledgements. Thanks are due to the National Science and Research Council of Canada and the Killam Foundation for financial support.

Chapter I

Interval Arithmetic

Suppose $p(x) = a_0 + a_1 x + \ldots + a_n x^n$ is a polynomial and suppose we wish to compute an inclusion for the range of values of $p(x)$ over an interval X. If we compute $p(X) = a_0 + a_1 X + \ldots + a_n X^n$ using interval arithmetic then the interval $p(X)$ will include the range as required. This shows that interval arithmetic is an important tool for our considerations. Some of the basic facts are therefore developed in this chapter.

1.1 THE TRICK OF INTERVAL ARITHMETIC

Interval arithmetic was introduced by Moore (1966). Here we briefly discuss the reasons for using interval arithmetic.

Present-day computers employ an arithmetic called fixed length floating point arithmetic or short floating point arithmetic. In this arithmetic real numbers are approximated by a subset of the real numbers called the machine representable numbers (or short machine numbers). Because of this representation two types of errors are generated. The first type of error occurs when a real valued input data item is approximated by a machine number. The second type of error is caused by intermediate results being approximated by machine numbers.

Interval arithmetic provides a tool for estimating and controlling these errors *automatically*. Instead of approximating a real value x by a machine number, the real value x is approximated by an interval X having machine number upper and lower boundaries containing the usually unknown value x. The width of this interval may be used as measure for the quality of the approximation. The calculations therefore have to be executed using intervals instead of real numbers and the real arithmetic is replaced by interval arithmetic.

Let us consider an example. The real number 1/3 cannot be represented by a machine number. It may, however, be enclosed in the machine representable interval $A = [0.33, 0.34]$ if we assume that the machine numbers are representable by two-digit numbers (without exponent part).

If we now want to multiply 1/3 by a real number b which we know lies in $B = [-0.01, 0.02]$ then we seek the smallest interval X which

(a) contains $b/3$,
(b) depends only on the intervals A and B, and does not depend on 1/3 and b,
(c) has machine numbers as boundaries.

The realization of these requirements is accomplished by two steps:

(i) Operations for intervals are defined which satisfy (a) and (b),
(ii) the application of certain rounding procedures to these operations yields (c).

These two steps and the properties resulting from their application have been developed extensively during the last two decades, cf. Moore (1966, 1979), Nickel (1975), Alefeld–Herzberger (1974, 1983). Further applications of interval arithmetic may be found in Nickel (1980).

It should again be mentioned that only a few principles of theoretical interval arithmetic are used in this monograph, and that it is possible to program the formulas and results on computers which are not equipped with interval arithmetic procedures. In the latter case, however, the rounding errors may falsify the results.

Readers who are interested in how to implement interval arithmetic on a computer may consult, for instance, Kulisch–Miranker (1981).

1.2 INTERVAL ARITHMETIC OPERATIONS

Let I be the set of real compact intervals (these are the ones normally considered). Operations in I satisfying the requirements (a) and (b) of Section 1.1 are then defined by the expression

$$A * B = \{a*b: a \in A, b \in B\} \quad \text{for } A, B \in I \tag{1.1}$$

where the symbol $*$ stands for $+$, $-$, $\cdot$, and $/$, and where A/B is only defined if $0 \notin B$.

The expression (1.1) is motivated by the fact that the intervals A and B include some exact values, α respectively β, of the calculation. The values α and β are generally not known. The only information which is normally available consists of the including intervals A and B, where $\alpha \in A$, $\beta \in B$. From (1.1) it now follows that

$$\alpha * \beta \in A * B.$$

This means that the (generally unknown) sum, difference, product, and quotient of the two reals is contained in the sum, difference, product, or the quotient of the including intervals.

It is emphasized that the real and the corresponding interval operations are denoted by the same symbols in the sequel. So-called *point intervals*, that is intervals consisting of exactly one point

$$[a, a]$$

are denoted by a. The expressions $aA, a + A, A/a, (-1)A$, etc. for $a \in R, A \in I$ are therefore defined (R denotes the set of real numbers). The expression $(-1)A$ is written as $-A$ in the usual manner.

It is always assumed in interval analysis that the operations given above are really interval operations, that is, the result of applying the above operations is always a compact interval. A proof of this fact appeared in Moore (1962), but this reference is not readily available. The following lemma is therefore included here:

Lemma 1.1 *If $A, B \in I$, then $A * B \in I$, if $*$ represents any interval operation provided $A * B$ is defined.*

Proof Let $*$ be one of the interval operations. The restriction of $*$ to the Cartesian product $A \times B$ is then equivalent to a real binary operation, and therefore continuous. Since $A \times B$ is compact with respect to the natural topology of R, it follows that $A * B$ as image of $A \times B$ under $*$ is also compact, see for example, Wilansky (1970). Furthermore $A \times B$ is connected. The image of a connected set under a continuous function is also connected. This means that $A * B$ is connected and $A * B$ is therefore a compact interval since these are the only compact and connected subsets of R, cf. Wilansky (1970). □

Proofs avoiding topological theorems also exist. They are based on some properties of R relating to order and continuity.

Expression (1.1) does not present a practical way of implementing the operations since it is not represented by an arithmetic expression. The following formulas are therefore used for practical calculations.

Lemma 1.2 *The definition of the interval operations is equivalent to*

$$[a, b] + [c, d] = [a + c, b + d],$$

$$[a, b] - [c, d] = [a - d, b - c],$$

$$[a, b][c, d] = [\min(ac, ad, bc, bd), \max(ac, ad, bc, bd)],$$

$$[a, b]/[c, d] = [a, b][1/d, 1/c] \text{ if } 0 \notin [c, d].$$

Proof. The proof that the above formulas result in the same results as those given by (1.1) is not difficult. One only has to show that any point of a result $[a, b] * [c, d]$ of the left side is also a point of the interval on the right side, and conversely. The execution of the proof is tedious mainly because of the occurrence of four possible cases. A fairly simple proof may be found in Ris (1972). □

Lemma 1.2 shows that subtraction and division in I are not the inverse operations of addition or multiplication as is the case in R. For example,

$$[0, 1] - [0, 1] = [-1, 1],$$
$$[1, 2]/[1, 2] = [1/2, 2].$$

This property is one of the main differences between interval arithmetic and real arithmetic. Another main difference is given by the so-called *subdistributive law*,

$$A(B + C) \subseteq AB + AC \quad \text{for } A, B, C \in I. \tag{1.2}$$

The proof of (1.2) is evident, from the fact that any point $x \in A(B + C)$ is of the form $x = a(b + c)$ for some $a \in A$, $b \in B$, and $c \in C$. This means that $x = ab + ac \in AB + AC$, from (1.1). It is not possible to prove the converse since a point $y \in AB + AC$ may only be represented by $y = a_1 b + a_2 c$ with $a_1, a_2 \in A$, $b \in B$, and $c \in C$ and since it is not always possible to find an $a \in A$ such that $y = a(b + c)$, i.e. $y \in A(B + C)$.

Example 1.1 $[0, 1][1 - 1] = 0$,

$$[0, 1]1 - [0, 1]1 = [-1, 1].$$

The distributive law is valid in some special cases.

Lemma 1.3 *For $a \in R$ and $B, C \in I$ we get*

$$a(B + C) = aB + aC. \qquad \square$$

The proof is trivial using Lemma 1.1. A complete list of conditions which lead to distributivity is given in Ratschek (1971) and Spaniol (1970).

In the following lemma we give some further properties of interval arithmetic.

Lemma 1.4 (i) *Addition and multiplication are commutative and associative.*

(ii) *For $A, B, C, D \in I$ and any interval operation $*$ it follows that $A \subseteq B$, $C \subseteq D$ implies $A * C \subseteq B * D$, if the operations are defined* (inclusion isotonicity *of the interval operations*). □

The proof of this lemma is evident using (1.1).

Let I^m be the set of all interval vectors $A = (A_1, ..., A_m)$, $A_i \in I$, and $o = (0, ..., 0)$. Interval vectors are usually interpreted as right parallelepipeds $A_1 \times ... \times A_m$. They will also be called *intervals* when it is clear from the context whether real intervals or interval vectors are intended. Vector operations are defined in the usual manner by

$$\alpha(A_1,\ldots, A_m) = (\alpha A_1,\ldots, \alpha A_m) \quad \text{if } \alpha \in R, A_i \in I,$$
$$(A_1,\ldots, A_m) \pm (B_1,\ldots, B_m) = (A_1 \pm B_1,\ldots, A_m \pm B_m)$$
$$\text{if } A_i, B_i \in I.$$

The inner product of I^m is denoted by

$$A \cdot B = A_1 B_1 + \ldots + A_m B_m.$$

Further relations on I^m are to be understood componentwise, for example $A \subseteq B$ or $a \in A$ for $A, B \in I^m$, $a \in R^m$.

If $D \subseteq R^m$, then $I(D) \subseteq I^m$ shall denote the set of all m-dimensional intervals $X \subseteq D$.

The *width* of an interval $A = [a, b]$ is defined by

$$w(A) = b - a.$$

The width of an interval vector $A = (A_1,\ldots, A_m) \in I^m$ is defined by

$$w(A) = \max \{w(A_i): i = 1,\ldots, m\}.$$

We list some properties of the width, cf. for example, Alefeld–Herzberger (1974, 1983): If $A, B \in I^m$, $C, D \in I$, $a \in R$, then

$$\left. \begin{array}{l} A \subseteq B \text{ implies } w(A) \leq w(B), \\ w(C \pm D) = w(C) + w(D), \\ w(aB) = |a|w(B). \end{array} \right\} \qquad (1.3)$$

These statements follow directly from the definitions. The *midpoint* of an interval $A = [a, b]$ is defined as

$$m(A) = (a + b)/2,$$

and the midpoint of an interval vector $A = (A_1,\ldots, A_m) \in I^m$ as

$$m(A) = (m(A_1),\ldots, m(A_m)).$$

The following important property follows directly from the definition of the midpoint,

$$m(A \pm B) = m(A) \pm m(B) \quad \text{if } A, B, \in I^m. \qquad (1.4)$$

The interested reader may find further properties of interval arithmetic in Alefeld–Herzberger (1974, 1983), Moore (1966, 1979), Nickel (1975, 1980).

1.3 NORMS AND METRICS IN INTERVAL SPACES

One could hardly conceive of the existence of an analysis not using the tools of norms and metrics. These tools are also essential for interval analysis. In this section we therefore introduce the basic facts on norms and metrics in interval spaces. (The reader may delay studying this section until the contents are needed.)

Let R^+ be the set of non-negative real numbers.

Definition 1.1 A function $\| \ \|: I^m \to R^+$ is called a *norm* on I^m if

1. $\|A\| = 0$ iff $A = o$,
2. $\|\alpha A\| = |\alpha| \, \|A\|$ if $\alpha \in R$,
3. $\|A + B\| \leq \|A\| + \|B\|$.

The norm is called *inclusion isotone* if

$$A \subseteq B \text{ implies } \|A\| \leq \|B\|.$$

We will only introduce one norm on I^m since all continuous norms on I^m are equivalent in the usual sense of analysis, as shown in the Appendix.

The expression

$$|A| = \max\{|a|, |b|\} \tag{1.5}$$

is called the *absolute value* (or *modulus* or *maximum norm*) of $A = [a, b]$. It can easily be checked that the absolute value is an inclusion isotone norm on I. If $A = (A_1, \ldots, A_m) \in I^m$ then

$$\|A\| = \max\{|A_i|: i = 1, \ldots, m\}$$

is called *maximum norm* of A. The maximum norm is also an inclusion isotone norm on I^m. In the sequel, the symbol $\| \ \|$ will always denote the maximum norm on I^m.

A number of properties for the absolute value have been verified in for example Alefeld–Herzberger (1974, 1983) or in Moore (1979). If $A, B \in I$ then we list a few of the properties:

$$\left.\begin{array}{l} |AB| = |A|\,|B|, \\ |A| \leq w(A) \text{ if } 0 \in A, \\ w(AB) \leq 2|A|, \\ w(AB) \leq w(A)|B| + |A|w(B), \\ w(1/A) \leq |1/A|^2 w(A) \quad \text{if } 0 \notin A. \end{array}\right\} \tag{1.6}$$

The proofs for these properties are found in the above references. Furthermore, if $A \in I^m$ and if $\| \ \|$ also denotes the maximum norm on R^m,

then
$$w(A) = \max\{\|a - b\|: a, b \in A\}. \tag{1.7}$$

Letting $x, y \in R$, we define (after Sunaga, 1958)
$$x \vee y = \begin{cases} [x, y] & \text{if } x \leq y, \\ [y, x] & \text{if } y \leq x. \end{cases}$$

We see that $x \vee y$ is the interval spanned by the reals x and y, that is, the smallest interval that contains x and y. It exists since the intersection of compact intervals is either a compact interval or the empty set. If $x = (x_1, \ldots, x_m)$, $y = (y_1, \ldots, y_m) \in R^m$, then we define
$$x \vee y = (x_1 \vee y_1, \ldots, x_m \vee y_m) \in I^m. \tag{1.8}$$

Clearly, $x \vee y$ is the interval that is spanned by the vectors x and y. We recognize the following relation:
$$w(x \vee y) = \|x - y\| \quad \text{for } x, y \in R^m.$$

Furthermore, if $X = (X_1, \ldots, X_m)$, $Y = (Y_1, \ldots, Y_m) \in I^m$ then the smallest interval $Z \in I^m$ that contains X and Y is denoted by $X \vee Y$ and is equal to $(X_1 \vee Y_1, \ldots, X_m \vee Y_m)$.

Although the definition of a metric is known and independent of special algebraic structures such as interval spaces we still mention it in order to be complete:

Definition 1.2 A function $\rho: I^m \times I^m \to R^+$ is called a *metric* on I^m if

1. $\rho(A, B) = 0$ iff $A = B$,
2. $\rho(A, B) = \rho(B, A)$,
3. $\rho(A, B) \leq \rho(A, C) + \rho(B, C)$,

holds for all $A, B, C \in I^m$. A metric ρ is called *homogeneous* if
$$\rho(aA, aB) = |a| \rho(A, B) \quad \text{for } a \in R, A, B \in I^m.$$

A metric ρ is called *translation invariant* if
$$\rho(A + C, B + C) = \rho(A, B) \quad \text{for } A, B, C \in I^m.$$

A metric ρ is called *chain inclusion isotone* (since the assumption is a chain) if
$$A \subseteq B \subseteq C \text{ implies } \rho(A, B) \leq \rho(A, C) \quad \text{for } A, B, C \in I^m$$
where the inclusion of interval vectors is understood componentwise.

The most important metric in interval arithmetic is the *Hausdorff metric* $|.,.|$ in I^m, defined as follows:
$$|[a, b], [c, d]| = \max\{|a - c|, |b - d|\},$$

and if $A = (A_1,\ldots, A_m)$, $B = (B_1,\ldots, B_m) \in I^m$,

$$|A, B| = \max\{|A_i, B_i|: i = 1,\ldots, m\}.$$

It can easily be shown that the Hausdorff metric is homogeneous, translation invariant, and chain inclusion isotone. Furthermore, the Hausdorff metric induces the maximum norm, that is,

$$|A, o| = \|A\| \qquad \text{for } A \in I^m. \tag{1.9}$$

Warning: Since subtraction in I and I^m is not the inverse operation of addition as is the case in vector spaces, the relation $|A, B| = \|B - A\|$ is not valid. For a discussion of this phenomenon, see Ratschek (1975).

Some properties of the Hausdorff metric in I are (see, for instance, Alefeld–Herzberger (1974, 1983) or Krawczyk–Nickel (1982)): If $A, B, C, D, A_i, B_i \in I (i = 1,\ldots, m)$ then

$$\left.\begin{array}{l} |AB, AC| \leq |A|\,|B, C|, \\ w(B) - w(A) \leq 2|A, B| \leq 2[w(B) - w(A)] \text{ if } A \subseteq B, \end{array}\right\} \tag{1.10}$$

$$|A + C, B + D| \leq |A, B| + |C, D|, \tag{1.11}$$

$$\left|\sum_{i=1}^{m} A_i, \sum_{i=1}^{m} B_i\right| \leq \sum_{i=1}^{m} |A_i, B_i|. \tag{1.12}$$

Formula (1.12) is proven by applying (1.11) and natural induction. Formula (1.11) is proven using the triangle inequality and the translation invariance of the Hausdorff metric:

$$|A + C, B + D| \leq |A + C, B + C| + |B + C, B + D| = |A, B| + |C, D|.$$

A further important formula needed is

$$|A \cdot B, A \cdot C| \leq m\|A\|\,|B, C| \qquad \text{for } A, B, C \in I^m. \tag{1.13}$$

It is only necessary to use (1.12) and (1.10) for the verification of this formula:

$$|A \cdot B, A \cdot C| \leq \sum_{i=1}^{m} |A_i B_i, A_i C_i| \leq \sum_{i=1}^{m} |A_i|\,|B_i, C_i|$$

$$\leq \sum_{i=1}^{m} \|A\|\,|B, C| = m\|A\|\,|B, C|$$

where A_i, B_i, C_i denote the components of $A, B,$ and C.

Lemma 1.5 (*Krawczyk–Nickel*, 1982). *If $A, B, C, D \in I$ and if $A \supseteq B \supseteq C \cap D \neq \emptyset$ then*

$$|A, B| \leq \max\{|A, C|, |A, D|\}.$$

Proof First we verify a special case of the lemma, that is

$$C \cap D = 0.$$

If in addition $C = 0$ or $D = 0$, the assertion follows directly from the definition of the Hausdorff metrics. Therefore we can assume that neither C nor D is 0 and that, without restricting the generality, C is to the left of D, that is,

$$C = [c, 0], \; c \leq 0,$$
$$D = [0, d], \; d \geq 0.$$

The remaining assumptions lead to the notation

$$A = [a_1, a_2], \; B = [b_1, b_2], \; \text{and} \; a_1 \leq b_1 \leq 0 \leq b_2 \leq a_2,$$

and the following inequalities:

$$|a_1 - b_1| \leq |a_1| \leq \max\{|a_1 - 0|, |a_2 - d|\} = |A, D|,$$
$$|a_2 - b_2| \leq a_2 \leq \max\{|a_2 - 0|, |a_1 - c|\} = |A, C|.$$

The assertion is evident when using the definition $|A, B| = \max\{|a_1 - b_1|, |a_2 - b_2|\}$.

The proof of the general case is reduced to the special case. It is only necessary to remember that, given intervals $U = [u_1, u_2] \subseteq V = [v_1, v_2]$, there exists exactly one interval $Z = [v_1 - u_1, v_2 - u_2]$ satisfying the equation

$$U + Z = V.$$

Let now intervals A, B, C, D be given as required by the lemma. Then there exist intervals A', B', C', and D' satisfying

$$A = A' + C \cap D, \; B = B' + C \cap D, \; C = C' + C \cap D,$$
$$D = D' + C \cap D.$$

It is easy to check that the intervals A', B', C', and D' satisfy the assumptions of the proven special case, i.e.,

$$A' \supseteq B' \supseteq C' \cup D' = 0.$$

Therefore we get the inequality

$$|A', B'| \leq \max\{|A', C'|, |A', D'|\},$$

which is equivalent to the asserted inequality by translation invariance. □

1.4 INCLUSION OF THE RANGE — A COMBINATORIAL PROBLEM?

Let us now investigate the set theoretic background for finding interval

Sec. 1.4] Inclusion of the Range — A Combinatorial Problem?

expressions that include the range of a function f in the same manner as the centred form will do it. The *range* of f over an interval X (in one or several dimensions) is denoted by

$$\tilde{f}(X) = \{f(x): x \in X\}.$$

What can now be done with a rational function f in order to get a centred form inclusion for $\tilde{f}(X)$? The answer is very easy: It is only necessary to write f in a certain manner, to replace the variable by the domain X, and to evaluate this expression using interval arithmetic operations.

In general, an inclusion of the range is obtained if f is written in a completely arbitrary form and the variable x is replaced by the domain X and then evaluated using interval arithmetic. For example, if

$$f(x) = x - x^2$$

is defined over the domain $X = [0, 2]$, then f may be written as

$$f(x) = x(1 - x)$$

or as

$$f(x) = c - c^2 + (1 - 2c)(x - c) - (x - c)^2 \text{ for some } c \in R.$$

The last representation is nothing but the Taylor expansion of f at the point c. The range of f over X is

$$\tilde{f}(X) = [-2, 1/4].$$

If we replace x by X in the above expressions and use simple interval arithmetic (that is, a power X^n is calculated as $X \cdot X \cdot \ldots \cdot X$ (n times), see Section 2.3) then the following inclusions of $\tilde{f}(X)$ are obtained:

$$X - X^2 = [-4, 2],$$
$$X(1 - X) = [-2, 2],$$

$$c - c^2 + (1 - 2c)(X - c) - (X - c)^2 = \begin{cases} [-4, 2] & \text{if } c = 0, \\ [-2, 2] & \text{if } c = 1, \\ [-2, 2] & \text{if } c = 1/2, \\ [-2, 3/2] & \text{if } c = 3/4. \end{cases}$$

From (1.1) it follows that the resulting intervals are always inclusions of $\tilde{f}(X)$ (see also Theorem 1.1). For example, if $x \in X$ then

$$f(x) \in c - c^2 + (1 - 2c)(X - c) - (X - c)^2.$$

Since the interval on the right side does not depend on x it follows that it contains all such $f(x)$ and therefore $\tilde{f}(X)$.

The question now arises how to write a function or how to choose an expression for f such that the replacement of x by X yields the best possible or at least a reasonable inclusion. Since there are infinitely many possibilities of writing the function f we cannot say that the search for an appropriate expression is a finite combinatorial problem. Also, we cannot say whether there exists a solution to this question at all. We can only prove that there is no better inclusion than some specified centred form if in the competition of all possible expressions for f a certain computational complexity is given, see Chapter V.

From the above discussion it clearly follows that it is necessary to both distinguish between a function and its defining arithmetic expression as well as to have some precise notation for the process of replacing x by X. The above heuristic discussion may then be made more precise.

Let $f(x)$ be any real *arithmetic expression* (abbreviated, an *expression*) in the real variable x or in the vector variable $x = (x_1, \ldots, x_m) \in R^m$. That means that $f(x)$ is a finite meaningful string of symbols in the sense of arithmetic consisting of

> the variable (or its components),
> real numbers (coefficients),
> the four arithmetic operations, and
> parentheses, brackets, etc.

This means that $x - xx$ is an expression in the variable $x \in R$, or abbreviated, $x - x^2$, if the common power notation is used. Furthermore, $x_1 - x_2 x_3$ or $3 - 4/(x_1 x_2 x_3)$ are expressions in the variable $x = (x_1, x_2, x_3) \in R^3$. The symbol strings $3 - (x$ or $+4$ or $\sin x$ are not expressions, however, since they are not meaningful strings in the above sense.

Two expressions are said to be *equal* if the corresponding strings are equal (and not the resulting functions). This means that an expression $f(x)$ is only then equal to $x - x^2$ if it is $x - x^2$ itself. The expressions $x(1 - x)$ and $x - x^2$ are not equal since the strings are not equal. Each arithmetic expression $f(x)$ clearly defines a rational function, and this expression is called the *underlying* or *defining expression* (or, abbreviated the *expression*) for this function.

Each rational function can be described by infinitely many expressions. The function $f(x) = x - x^2$ for example has among others the following underlying expressions:

$$x - x^2 + nx - nx \quad \text{for each integer } n.$$

An *interval expression* (usually abbreviated to an *expression*) is defined analogously to an arithmetic expression. In the text of the definition one has only to replace 'variable' by 'interval variable' and 'real numbers' by 'intervals'. The operations are clearly then the interval operations.

If $f(x)$ is an expression and X an interval of the same dimension as x, then the (interval-) expression which is obtained by replacing each occurrence of x in $f(x)$ by X is denoted by $f(X)$. The expression $f(X)$ is then called the *natural interval extension* of $f(x)$ on X, see Moore (1979). The function which is defined by the assignments $X \to f(X)$ is called the *natural interval extension of $f(x)$*.

For simplicity of notation we also allow $f(x)$ or f to denote the rational function defined by the expression $f(x)$, and we speak usually about the *natural interval extension of the function $f(x)$* on X if there is no doubt or confusion regarding the underlying expression. We also denote by $f(X)$ the interval obtained as the value of the calculation implied by the expression $f(X)$ if there are no disallowed divisions. In this case the interval $f(X)$ is called the *value* of the interval extension.

The following theorem due to Moore (1966) forms the basis for the theory of inclusions of ranges.

Theorem 1.1 *Let $D \subseteq R^m$, $X \in I(D)$, and let $f(x)$ be an arithmetic expression in the variable $x \in D$. If the interval extension $f(X)$ is defined, then*

$$\bar{f}(X) \subseteq f(X).$$

Proof Let $a \in \bar{f}(X)$. Then there exists an argument $y \in X$ such that $f(y) = a$. The application of Lemma 1.4 yields $a = f(y) \in f(X)$. □

If f is a rational function, then $f(x)$ is undefined as an expression, since the underlying expressions are not unique. When we define a rational function, however, it is usually written down as an 'expression' which then shall be seen as *the* current defining expression of the function. For example, let $p(x) = 1 + 3x - 5x^2$ be a polynomial. By the expression $p(x)$ we then imply that the expression $1 + 3x - 5x^2$ is considered.

We will also use the usual abbreviations of arithmetic when we write expressions. For example

$$x^n, \sum_{i=0}^{n} a_i x^i$$

where x^0 means 1, etc.

We are certain that our definition of an expression is sufficient for the purposes of this book. The reader who feels that this definition is not precise enough may choose a recursive definition and may orient himself by looking at the definitions of a well-formed formula in logic, cf. Mendelson (1964) or a polynomial symbol in universal algebra, cf. Grätzer (1979).

1.5 INCLUSION ISOTONICITY AND LIPSCHITZ CONDITIONS

In this section we will consider some connections between real and interval functions. These conditions enable us to carry certain properties of real functions over to interval functions or to guarantee similar or other properties of interval functions. Most of the material of this section was first given in Moore (1966, 1979).

Let $D \subseteq I^m$ and let $F:I(D) \to I^k$ be an interval function. F is said to be *inclusion isotone* if

$$X \subseteq Y \text{ implies } F(X) \subseteq F(Y) \quad \text{for all } X, Y \in I(D).$$

Lemma 1.4 states that the four interval operations are inclusion isotone. The *range function* $\bar{f}:I(D) \to I^k$ of a function $f: D \to R^k$ which is defined by $\bar{f}(Y) = \{f(y): y \in Y\}$ is furthermore inclusion isotone. Many interval functions are not inclusion isotone, however, for example, the interval function $F: I \to I$, defined by

$$F(X) = m(X) + (X - m(X))/2$$

is not inclusion isotone, see Moore (1979). We have $F([0, 2]) = [1/2, 3/2]$, and $F([0, 1]) = [1/4, 3/4]$ and clearly $F([0, 1])$ is not contained in $F([0, 2])$.

It is important to consider inclusion isotonicity since various properties of centred forms are connected to inclusion isotonicity as seen in the sequel. The following lemma shows that inclusion isotonicity is a very 'natural' property that is valid in the large and frequently used class of natural interval extensions.

Lemma 1.6 *If f is an arithmetic expression in one or several variables then the natural interval extension of f is inclusion isotone.*

Proof The four interval arithmetic operations are inclusion isotone. The assertion follows by complete induction since a rational function consists only of finitely many applications of the arithmetic operations. □

It is clear that the concatenation of inclusion isotone functions is inclusion isotone.

In order to handle Lipschitz conditions it is necessary to say something about the manipulation of limits and the topology used in I^m. The natural way of defining limiting operations is via the $2m$ endpoints of the m coordinate intervals or the $2m$ related boundary functions, cf. Apostolatos–Kulisch (1967).

For example, let $(Y_n)_{n=0}^{\infty}$ be a sequence of intervals of I^m and $Y_n = (Y_{n1},\ldots, Y_{nm})$, $Y_{ni} = [y_{ni}, z_{ni}]$, $i = 1,\ldots, m$. This sequence is said to *converge to* an interval $\check{Y} = (\check{Y}_1,\ldots, \check{Y}_m)$, $\check{Y}_i = [\check{y}_i, \check{z}_i]$, $i = 1,\ldots, m$, iff for each $i = 1,\ldots, m$ the endpoint sequences $(y_{ni})_{n=0}^{\infty}$ and $(z_{ni})_{n=0}^{\infty}$ converge to $\check{y}_i$ and $\check{z}_i$ respectively. This means that the topology on I^m is defined in such a manner that the mapping $i: I^m \to R^{2m}$, defined by

$$([y_1, z_1],\ldots, [y_m, z_m]) \to (y_1, z_1,\ldots, y_m, z_m), \tag{1.14}$$

is a topological embedding with respect to the natural topology on R^{2m}. This topology can also be defined via the Hausdorff metric on I^m, see, for example, Moore (1979).

Let $D \subseteq I^m$, $Y \in I(D)$, and $F: I(D) \to I^k$. In the above topology F is said to be *continuous* at Y if $\lim_{n\to\infty} F(Y_n) = F(Y)$ for each sequence $(Y_n)_{n=0}^{\infty}$ converging to Y. If F is continuous at each $Y \in D$, then F is said to be *continuous* (in D). This definition of continuity is equivalent to the ε- δ-definition expressed by the Hausdorff metric, as is the case in analysis. A simple proof may be made using the fact that the definition of continuity involves the endpoints of the coordinate intervals. In this manner we get a componentwise reduction to known analogous relations in real analysis.

As an example we consider an interval valued polynomial $P(x) = \sum_{i=0}^{n} A_i x^i$, where $A_i = [a_i, b_i]$ and $x \in R^+$. Since the boundary functions of P are continuous in R^+ it follows that P is continuous. The boundary functions, defined by $P(x) = [g(x), h(x)]$, are $g(x) = \sum_{i=0}^{n} a_i x^i$ and $h(x) = \sum_{i=0}^{n} b_i x^i$.

We note, however, that such an easy definition cannot be given for the derivative concept, see Ratschek–Schröder (1971) and Markov (1977).

Lemma 1.7 *Addition, subtraction, multiplication, division, max, and min in I are continuous operations.*

Proof The limiting process that is necessary in the proof is separated into limiting processes for the left and right endpoints. In this manner the proof is reduced to the continuity of addition, subtraction, multiplication, division, max, and min in R. □

We now introduce a kind of a Lipschitz condition (Moore, 1979). This condition will be essential in our investigation of quadratic convergence in Section 3.3.

Definition 1.3 Let $D \subseteq R^m$ and $F: I(D) \to I^k$. Then F is called *Lipschitz* if there exists a real number K (*Lipschitz constant*) such that

$$w[F(Y)] \leq K w(Y) \quad \text{for all } Y \in I(D).$$

The properties of being continuous and Lipschitz are independent. The following examples are designed to demonstrate this independence.

Example 1.2 (Lipschitz condition does not imply continuity). Let δ be Dirichlet's jump function,

$$\delta(x) = 1 \text{ if } x > 0,$$
$$= 0 \text{ if } x \leq 0.$$

Let $F: I(X) \to I$ with $X = [0, 1]$ be defined by

$$F(Y) = Y + \delta(w(Y)).$$

Then by (1.3),

$$w[F(Y)] = w(Y),$$

and F is Lipschitz. However, F is not continuous at 0. The sequence $(Y_n)_{n=1}^{\infty}$, given by $Y_n = [0, 1/n]$ converges to 0, the sequence of the function values $F(Y_n) = [0, 1/n] + 1$ converges to 1, but $F(0) = 0$.

Example 1.3 (Continuity does not imply a Lipschitz condition). Let $X = [0, 1]$ and let the function $F: I(X) \to I$ be defined by

$$F(Y) = [0, \sqrt{w(Y)}].$$

Then, F is continuous but not Lipschitz in $I(X)$. The continuity follows from the continuity of the square-root function and the width function. Furthermore, a Lipschitz constant cannot exist, since the set of quotients

$$w[F(Y)]/w(Y) = \sqrt{w(Y)}/w(Y) = 1/\sqrt{w(Y)}, \quad w(Y) \neq 0,$$

is unbounded.

Lemma 1.8 *For an interval $X \in I^m$, the set $I(X)$ is compact in I^m.*

Proof The set $I(X)$ is compact iff each sequence in $I(X)$ has a convergent subsequence, see, for example, Wilansky (1970). Now, if $X = (X_1, \ldots, X_m)$ and if a sequence $(Y_n)_{n=0}^{\infty}$ is given in $I(X)$, where $Y_n = ([y_{n1}, z_{n1}], \ldots, [y_{nm}, z_{nm}])$ and if the image of Y_n under the homeomorphism i, see (1.14) is denoted by $\tilde{Y}_n = (y_{n1}, z_{n1}, \ldots, y_{nm}, z_{nm})$, then $(\tilde{Y}_n)_{n=0}^{\infty}$ is a sequence in the compact set $X_1 \times X_1 \times \ldots \times X_m \times X_m \in R^{2m}$ and has therefore a convergent subsequence, whose preimage is a convergent subsequence of $(Y_n)_{n=0}^{\infty}$. □

The following theorem shows that both the large class of natural interval extensions of rational functions, and also the class of those extensions which are combined with continuous Lipschitz functions as the midpoint function $Y \to m(Y)$ or the width function $Y \to w(Y)$ are Lipschitz.

Theorem 1.2 *Let $X \in I^m$ and let the functions $\varphi_1, \ldots, \varphi_k: I(X) \to I$ be continuous and Lipschitz. Assume that the function $F: I(X) \to I$ is constructed from the functions $\varphi_1, \ldots, \varphi_k$, interval constants, and a variable $Y \in I(X)$, using only the four interval operations. Then F is Lipschitz.*

Proof The basic idea of the proof is due to Moore (1979). In the proof complete induction is used with respect to the number of occurrences of the interval operations in F. With the exception of the start of the induction, each induction step can be executed by one of the following cases.

If $G, H: I(X) \to I$ are continuous Lipschitz functions with Lipschitz constants K and L respectively then each of the following functions are

continuous and Lipschitz:
(a) $G \pm H$, and this function has a Lipschitz constant which is $K + L$,
(b) GH, and this function has a Lipschitz constant which is given by $L \max \{|G(Y)|: Y \in I(X)\} + K \max \{|H(Y)|: Y \in I(X)\}$,
(c) aG for $a \in R$, these functions have Lipschitz constants $|a|K$,
(d) $1/G$, if defined and this function has a Lipschitz constant which is $K \max \{|1/G(Y)|^2: Y \in I(X)\}$.

These four cases can easily be verified:
(a) By (1.3),
$$w[G \pm H)(Y)] = w[G(Y)] + w[H(Y)] \leq (K + L)w(Y).$$
(b) By (1.6),
$$w[(GH)(Y)] \leq |G(Y)|w[H(Y)] + |H(Y)| w[G(Y)] \leq |G(Y)|L + |H(Y)|K.$$

The last expression can be estimated upwards as asserted. The maxima exist since $I(X)$ is compact and both G, H and the norm are continuous.
(c) By (1.3),
$$w[aG(Y)] = |a|w[G(Y)] \leq |a|Kw(Y).$$
(d) By (1.6),
$$w[1/G(Y)] \leq |1/G(Y)|^2 w[G(Y)] \leq |1/G(Y)|^2 Kw(Y).$$

The maximum of the Lipschitz constant exists due to the compactness and the continuity. □

Remark 1.1 For simplicity, Theorem 1.2 is formulated only for simple interval valued functions and not for interval vector valued functions. It is naturally also valid in this last mentioned case.

Corollary 1.1 *If F is a natural interval extension of a rational function, then a Lipschitz constant for F can be computed explicitly in finitely many steps.*

Proof The analytical steps of the proof of Theorem 1.2 consist of the determinations of the various maxima. In the absence of the functions $\varphi_1, \ldots, \varphi_k$, these determinations can be replaced by arithmetic or logical steps,
$$\max \{|G(Y)|: Y \in I(X)\} = |G(X)|,$$
$$\max \{|1/G(Y)|^2: Y \in I(X)\} = |1/G(X)|^2,$$
etc., because the norm and the initial operations are inclusion isotone. □

Example 1.4 Let $X = [0, 1]$ and let $F: I(X)^4 \to I$ be given by
$$F(Y) = Y_1 Y_2 + Y_3/(1 + Y_4) \quad \text{for } Y \in I(X)^4.$$
Then by (1.3),
$$w[F(Y)] = w(Y_1 Y_2) + w[Y_3/(1 + Y_4)].$$

Using Corollary 1.1 and the formulas of the proof of Theorem 1.2, we get

$$w(Y_1 Y_2) \leq |X| w(Y_2) + |X| w(Y_1) \leq 2|X| w(Y)$$

and

$$\begin{aligned} w[Y_3/(1 + Y_4)] &\leq |X| w[1/(1 + Y_4)] + 1/(1 + X) w(Y_3) \\ &\leq |X| |1/(1 + X)|^2 w(Y_4) + |1/(1 + X)| w(Y_3) \\ &\leq (|X| |1/(1 + X)|^2 + |1/(1 + X)|) w(Y). \end{aligned}$$

Since $|X| = 1$ and $|1/(1 + X)| = |[1/2, 1]| = 1$, the number 4 is a Lipschitz constant for F, and clearly, F is Lipschitz,

$$w[F(Y)] \leq 4w(Y) \quad \text{for all } Y \in I(X).$$

A trivial but important lemma is the following:

Lemma 1.9 *If F and G are interval functions that satisfy the Lipschitz condition and the concatenation $G \circ F$ is defined, then $G \circ F$ is Lipschitz.*

Proof Clearly, if K and L are the Lipschitz constants of F and G, then

$$w[G \circ F(Y)] = w(G[F(Y)]) \leq L\, w[F(Y)] \leq LK w(Y). \quad \square$$

The significance of the following theorem is its applicability to real (and not only to rational) functions.

Theorem 1.3 *Let $X \in I^m$ and let the function $\bar{f}\colon X \to R$ be Lipschitz in the usual sense of analysis, that is, there exists a constant K such that*

$$|f(x) - f(y)| \leq K \|x - y\| \quad \text{for all } x, y \in X. \tag{1.15}$$

Then the range function $\bar{f}\colon I(X) \to R$, given by $Y \to \bar{f}(Y)$, satisfies the Lipschitz condition.

Proof It is shown that

$$w[\bar{f}(Y)] \leq K w(Y) \quad \text{for all } Y \in I(X)$$

holds. Let $Y \in I(X)$. Since Y is compact and since f is necessarily continuous, there exist vectors $x = (x_1, \ldots, x_m)$, $y = (y_1, \ldots, y_m) \in Y$ such that $\bar{f}(Y) = [f(x), f(y)]$. By (1.15), it follows that

$$\begin{aligned} w[\bar{f}(Y)] &= f(y) - f(x) \leq K \|y - x\| \\ &= K \max\{|y_i - x_i|: i = 1, \ldots, m\} \leq K w(Y). \end{aligned}$$

Although Theorem 1.3 is formulated with the maximum norm it holds for any norm since each two norms are equivalent in R^m.

Example 1.5 If $X = [x_1, x_2] \in I$ and if $Y = [y, z] \in I(X)$ is an interval variable then the following range functions are Lipschitz, cf. Moore (1979):

$$\overline{e^Y} = [e^y, e^z],$$
$$\overline{Y^{1/2}} = [y^{1/2}, z^{1/2}] \text{ if } x_1 > 0,$$
$$\overline{\ln Y} = [\ln y, \ln z] \text{ if } x_1 > 0,$$
$$\overline{\sin Y} = [\sin y, \sin z] \text{ if } X \subseteq [-1, 1]\,\pi/2,$$
$$\overline{\sin Y} = [\min \{\sin x: x \in Y\}, \max \{\sin x: x \in Y\}],$$

etc.

CHAPTER II

The standard centred form

Among the most important inclusions of the ranges of functions are the so-called centred forms, first suggested by Moore (1966). Explicit formulas for these forms were found by Hansen (1969) for polynomials and by Ratschek (1978) for rational functions. They will now be called the standard centred forms in the sequel. These forms will therefore be developed and investigated in detail in this chapter. Recursively defined centred forms were recently introduced by Krawczyk (1983). These forms of lower complexity than the standard centred forms, will also be discussed in this chapter.

2.1 INTRODUCTION

The concept of a centred form is due to Moore (1966). When he computed the range of specific functions he noticed that he got better results on the average if he developed the functions in a certain manner around the centre of the domain of the function. Moore discussed this phenomenon in some detail and he also compared these forms with other possibilities for evaluating the range of a function. His observations were later confirmed by the work of Goldstein–Richman (1973). The original definition of Moore was: Let f be a rational function in one real variable x and let X be an interval contained in the domain of f. Furthermore, let c be the midpoint of X, that is, $c = m(X)$. It then follows that there exists an arithmetic expression $s(y)$ such that

$$f(x) = f(c) + s(x - c) \tag{2.1}$$

is valid. It is assumed that the expression $s(y)$ is as simple as possible. If we replace the variable x by the interval X and if we use interval arithmetic instead of real arithmetic in the evaluation of the expression we obtain the *centred form* of f on X, denoted by

$$F(X) = f(c) + s(X - c).$$

This centred form had the property that

$$F(X) \supseteq \bar{f}(X)$$

where $\bar{f}(X) = \{f(x): x \in X\}$ denotes the range of f over X. Moore wrote (1966, p. 45) that his definition of the centred form was 'rather vague', and he hoped that a 'more precise and elegant presentation will be found'.

An explicit and useful definition of the centred form was not discovered immediately. In the case of polynomials, Hansen (1969) noticed that a useful centred form was the natural interval extension of the Taylor expansion of the polynomial developed at the point c. He also noticed that the function s must be written in the form

$$s(x) = (x - c)g(x - c)$$

in order to get the so-called quadratic convergence which is a measure of how quickly arbitrarily good inclusions can be computed, see for example, the end of this section as well as Chapter IV and Section 3.3. The function f is then written as

$$f(x) = f(c) + (x - c)g(x - c), \tag{2.2}$$

and the centred form of f on X is

$$F(X) = f(c) + (X - c)g(X - c). \tag{2.3}$$

It also turned out that the underlying expression for g was not of importance.

Explicit formulas for centred forms for arbitrary rational functions in one and several variables were given in Ratschek (1978, 1980a). It was also shown in Ratschek (1978, 1980a) that better approximations of the range can in general be found, if the condition that the expression $s(y)$ should be as simple as possible is dropped. Krawczyk's (1983) recursively defined centred forms are also of importance because of their low complexity.

The first attempt to define centred forms for real functions was undertaken by Alefeld–Herzberger (1974). A very convenient and useful version may also be found in Krawczyk–Nickel (1982). This version will be developed in Chapter III. Centred forms for interval valued polynomials are defined in Rokne (1981), and centred forms for rational functions over the complex plane are studied in Rokne–Wu (1982, 1983). One of the most recent results is due to Krawczyk (1982). His results show that the so-called Krawczyk operator may be described using centred forms for operators.

Many authors have investigated the properties of the centred form for the case that c is not necessarily the midpoint of the domain of f, see Chuba–Miller (1972), Miller (1972), Alefeld–Herzberger (1974, 1983), Krawczyk–Nickel (1982), Moore (1966). We discuss this point in Section 2.3.

The so-called *quadratic convergence* is closely connected to the treatment of the centred forms. The quadratic convergence is usually written as

$$w(F(X)) - w(\tilde{f}(X)) = O(w(X))^2,$$

if $w(X)$ converges to 0. This means that there exists a constant $\gamma \in R$ depending on f and on X such that

$$w(F(Y)) - w(\tilde{f}(Y)) \leq \gamma w(Y)^2 \quad \text{for } Y \in I(X).$$

Quadratic convergence indicates that it is appropriate to use centred forms since it shows that the inclusion of $\tilde{f}(X)$ can be improved 'quadratically', if the domain X becomes smaller or if X is subdivided, see Section 4.2. Centred form definitions are now in general only accepted if they have the property of quadratic convergence which guarantees an efficient improvement of the including estimation by subdivision methods.

In this chapter we discuss some classes of centred forms, the standard and Krawczyk's centred forms. If the reader is then familiar with these forms and their properties he may easier understand the axiomatic viewpoint that will be discussed in Chapter III. If the reader is only interested in concrete inclusions for the applications then the explicit representations of this chapter are more direct effective tools.

2.2 THE STANDARD CENTRED FORM FOR RATIONAL FUNCTIONS IN ONE VARIABLE

In this section we introduce the standard centred forms for functions in one variable since this case provides a clear view of the main ideas. The case of several variables is obscured by technical details. An example of this is the use of multi-indices necessary for a reasonable development of centred forms in several dimensions. These are treated in Section 2.5.

Let $f = p/q$ be a rational function of one real variable represented by the quotient of two polynomials p and q. Let n be the maximum of the degrees of p and q. For a real compact interval X contained in the domain of f and the midpoint $c = m(X)$ we define

$$t_i = p^{(i)}(c) - f(c)q^{(i)}(c) \quad \text{for } i = 1,\ldots,n. \tag{2.4}$$

Let also $H = X - c$.

Definition 2.1 The interval

$$F(X) = f(c) + \frac{t_1 H + \ldots + t_n H^n/n!}{q(c) + q'(c)H + \ldots + q^{(n)}(c)H^n/n!}$$

Rational Functions in one Variable

is called the *standard centred form* of f on X if 0 does not lie in the denominator.

It should be noted that although we stated in Section 1.4 that $f(X)$ would in general denote the natural interval extension of a particular expression we reserve the notation $F(X)$ for the standard centred form of f on X in this chapter.

We furthermore eschew the recent trend in the study of centred forms of being as general as possible and to admit any point of the domain of f as the point c. A critical discussion of our standpoint can be found in Remark 2.5. The reader may, however, disagree with us and join this line of investigation.

We now have to show that Definition 2.1 is reasonable and that $F(X)$ is indeed a centred form as shown in our initial discussion. This therefore implies that $F(X)$ should be quadratically convergent and that $F(X) \supseteq \tilde{f}(X)$. The quadratic convergence is shown in Section 4.2. Therefore we restrict ourselves to proving that

$$\tilde{f}(X) \subseteq F(X) \tag{2.5}$$

holds if $F(X)$ is defined. For this purpose we define

$$s(y) = \left(\sum_{i=1}^{n} t_i y^i/i! \right) / \left(\sum_{i=0}^{n} q^{(i)}(c) y^i/i! \right). \tag{2.6}$$

The standard centred form is then the natural interval extension of the expression $f(c) + s(y)$ to the interval H (that is, y is replaced by H). Equivalently, it may be considered to be the natural interval extension of the expression $f(c) + s(x - c)$ to X (that is, x is replaced by X). We only have to demonstrate that the expression $f(c) + s(x - c)$ defines f, that is, that $f(x)$ and $f(c) + s(x - c)$ are identical as functions. Then (2.5) is verified by Theorem 1.1.

We write f in the following form:

$$f(x) = \frac{p(x)}{q(x)} = \frac{p(x)q(c) + p(c)q(x) - p(c)q(x) + p(c)q(c) - p(c)q(c)}{q(x)q(c)}$$

$$= \frac{p(c)}{q(c)} + \frac{p(x) - p(c)}{q(x)} - f(c) \frac{q(x) - q(c)}{q(x)}$$

$$= f(c) + \frac{(p(x) - p(c)) - f(c)(q(x) - q(c))}{q(x)}$$

Writing the expressions $q(x)$, $p(x) - p(c)$ and $q(x) - q(c)$ using their Taylor expansions we obtain the desired expression for $f(c) + s(x - c)$ after an unimportant rearrangement. This proves (2.5).

Example 2.1 Let $f(x) = 1 - x + x^2$ and $X = [0, 2]$. For $c = 1$ we get $t_1 = 1$, $t_2 = 2$, and $f(c) = 1$, such that

$$f(x) = f(c) + t_1(x - c) + t_2(x - c)^2/2 = 1 + (x - c) + (x - c)^2,$$
$$F(X) = 1 + H + H^2 = 1 + [-1,1] + [-1,1]^2 = [-1,3].$$

The range of f over X is $\bar{f}(X) = [3/4, 3]$.

Example 2.2 Let the rational function $f(x) = (-3x^4 + 4x^3 - 2)/(4x^2 + 4x + 2)$ be evaluated over the interval $X = [0.8, 1.2]$. The range is first estimated by evaluating f at 100 equally spaced points in X obtaining $\bar{f}(X) \simeq [-0.1522, -0.09489]$ with $w[\bar{f}(X)] \simeq 0.05728$. (Evaluating at 1000 equally spaced points did not change the above results.) We furthermore obtained $f(X) = [-0.2314, 0.09856]$ with $w[f(X)] = 0.3299$. Estimating the range using the centred form resulted in $F(X) = [-0.1717, -0.02838]$ with $w[F(X)] = 0.1433$.

Example 2.3 The range of the rational function $f(x) = (3x^2 - x + 3.5)/(4x^4 + x^3 + 2)$ is to be estimated over the interval $X = [0.9, 1.1]$. Estimating the range as in the previous example we obtained $\bar{f}(X) \simeq [0.6958, 0.8942]$ with $w[\bar{f}(X)] \simeq 0.1984$ (again no difference between 100 and 1000 evaluations in the estimation). Furthermore, $f(X) = [0.5210, 1.195]$ such that $w[f(X)] = 0.6732$. Using the standard centred form we obtain $F(X) = [0.6141, 0.9573]$ with $w[F(X)] = 0.3432$.

If we use the techniques of extended power evaluation and of the nested form evaluation we obtain better approximations to $\bar{f}(X)$ than those obtained by computing $F(X)$. These improvements are of a technical nature and we refer the discussion of these improvements to Section 2.3 since the expressions become more involved and unclear. The symmetry of $s(X - c)$ is lost and the improvement vanishes quadratically as $w(X)$ tends to zero.

In order to determine the width of the centred form and in order to derive a criterion for the existence of the centred form we need the relations:

$$0 \notin a + [-v, v] \text{ is equivalent to } |a| > v \text{ for } a \in R, v \geq 0, \tag{2.7}$$

$$w\left(\frac{[-u,u]}{a + [-v,v]}\right) = \frac{2u}{|a| - v} \quad \text{for } u, v \geq 0, |a| > v. \tag{2.8}$$

The proof of (2.7) is trivial. Equation (2.8) can be verified by a simple rearrangement of the interval

$$\frac{[-u,u]}{a + [-v,v]} = \frac{[-u,u]}{[a - v, a + v]} = [-u,u]/(|a| - v).$$

In the following lemmas we let $z = w(X)/2$ such that

$$H = X - c = [-z, z]. \tag{2.9}$$

Lemma 2.1 *The standard centred form $F(X)$ for f on X is defined if and only if*

$$|q(c)| - |q'(c)|z - \ldots - |q^{(n)}(c)|z^n/n! > 0.$$

The *proof* follows directly from (2.7). □

Lemma 2.2 *The width of the standard centred form is given by*

$$w[F(X)] = 2 \frac{|t_1|z + \ldots + |t_n|z^n/n!}{|q(c)| - |q'(c)|z - \ldots - |q^{(n)}(c)|z^n/n!}$$

Proof The lemma is a direct consequence of the definition of the standard centred form and the relations (1.3) and (2.8). □

Remark 2.1 If f is already a polynomial, $f = p$, then the expression $f(c) + s(x - c)$ can be interpreted as the Taylor polynomial of p at c, cf. Hansen (1969b). The standard centred form is therefore

$$P(X) = p(c) + p'(c)H + \ldots + p^{(n)}(c)H^n/n!. \tag{2.10}$$

Remark 2.2 The denominator of the standard centred form of f is the standard centred form for q on X, see Remark 2.1. This means that $F(X)$ can be written as

$$F(X) = f(c) + \left(\sum_{i=1}^{n} t_i H^i/i! \right)/Q(X).$$

Remark 2.3 Considering the coefficients t_i of the standard centred form we almost recognize the Taylor coefficients of p and q. If

$$p_i(c) = p^{(i)}(c)/i! \quad \text{and} \quad q_i(c) = q^{(i)}(c)/i!,$$

for $i = 0, 1, \ldots, n$, denote the Taylor coefficients of p and q at c then the standard centred form can be written as

$$F(X) = f(c) + \frac{\sum_{i=1}^{n} [p_i(c) - f(c)q_i(c)]H^i}{\sum_{i=0}^{n} q_i(c)H^i}.$$

One should therefore use Moore's recursive technique of evaluating Taylor coefficients (Moore, 1966, 1979) in order to have an *effective* and *fast* method for calculating the coefficients t_i. It is only necessary to make a small change in this technique in order to compute the terms $p^{(i)}(c)$ and $q^{(i)}(c)$ of t_i.

Remark 2.4 If there is no interval arithmetic compiler available then there is a very simple procedure for computing the standard centred form. It is sufficient to note that each interval $A = [a, b]$ admits the representation

$$A = m(A) + [-1,1]w(A)/2 \qquad (2.11)$$

which easily can be verified by writing $m(A) = (a + b)/2$ and $w(A) = b - a$. The meaning of (2.11) is that an interval A is uniquely determined by its midpoint and its width:

$$\left.\begin{array}{l} a = m(A) - w(A)/2, \\ b = m(A) + w(A)/2. \end{array}\right\} \qquad (2.12)$$

This means that if we wish to calculate $F(X)$ we first apply Lemma 2.1. This tells us whether $F(X)$ exists or not. If it exists then we may evaluate the endpoints of $F(X)$ using (2.12), taking $f(c)$ to be the midpoint and computing the width $w[F(X)]$ as in Lemma 2.2. It should, however, be realized that the computation of $F(X)$ in this manner will not guarantee an inclusion of $\tilde{f}(X)$ due to round-off error contamination, although the chances are that the inclusion indeed is valid.

Remark 2.5 It is theoretically possible to choose any c from the domain of f in order to define $F(X)$, in the definition of the standard centred form. This would certainly constitute a more general definition. The current state of both theoretic and practical investigations indicates, however, that the centre of X should be chosen for c. The reasons are:

(i) Practical experience has shown that in the average, the centre is a very good value for c, see Section 2.3.

(ii) So far there has been no investigation made of the best possible choice of c depending on f and X.

(iii) If $c = m(X)$, then all the intervals $H = X - c$ and their powers are symmetric intervals. This leads to considerable computational and theoretical advantages, see Section 2.3. For instance, $f(c)$ is always the midpoint of $F(X)$. Because of the symmetry there also exist simple formulas for the width of $F(X)$ such that the quality of $F(X)$ can easily be estimated.

(iv) The evaluation of the standard centred form without using an interval arithmetic compiler causes no problems and no additional computational effort, see Remark 2.4.

(v) In the discussion of the general centred form it is known that the overestimation of the range $\tilde{f}(X)$ by the centred form can be improved by 50 per cent if $c = m(X)$, see Remark 3.2.

(vi) Almost all the operators which can be described by a generalized

2.3 SYMMETRIC INTERVALS — OR NOT?

Intervals of the form $[-a, a]$ for $a \geq 0$ are called *symmetric* intervals. They possess important computational and theoretic advantages which are the main reasons for choosing as c the midpoint of X in the definition of the centred form, see Remark 2.5. All the intervals $H,\ldots,H^n$ are now symmetric according to this definition.

Some of the facilities for calculating with symmetric intervals which are used when considering centred forms are:

(1) Adding or subtracting a symmetric interval $[-a, a]$ to an interval A preserves the midpoint,

$$m(A + [-a, a]) = m(A).$$

(2) If one factor of an interval product is symmetric then the product is symmetric.

(3) Symmetric intervals can be represented by only one parameter, for instance the right endpoint,

$$[-a, a] = [-1, 1]a \quad \text{for } a \geq 0.$$

Addition and multiplication of symmetric intervals can therefore be reduced to addition or multiplication of the parameters:

$$\left.\begin{array}{l} [-a, a] \pm [-b, b] = [-1, 1](a + b), \\ [-a, a][-b, b] = [-1, 1]ab \end{array}\right\} \text{ for } a, b \geq 0.$$

(4) If A is symmetric and $B = [c, d]$ then the following formulas hold:

$$\left.\begin{array}{l} AB = A|B|, \\ A/B = \begin{cases} A/c \text{ if } c > 0, \\ A/d \text{ if } d < 0, \end{cases} \\ w(AB) = w(A)|B| = 2|A|\,|B|. \end{array}\right\} \quad (2.13)$$

All these assertions can be verified easily. We notice that (2) follows directly from (4). The standard centred form can furthermore be computed without additional effort if the user does not have interval arithmetic software because of (3). The second formula of (4) was used for deriving formula (2.8). If (4) is compared to the multiplication and division rules (Lemma 1.2) then the large improvement in the computational complexity when using (4) should not be disregarded.

If one is willing to give up the advantages offered by the use of symmetric intervals, one may

(a) choose an arbitrary point of the domain of f as c,
(b) use extended power evaluation,
(c) use the nested form (also known as small Horner scheme) if $c \neq m(X)$.

As we know, (a) permits a more general definition of the centred form and leads sometimes to better (but frequently to worse) inclusions of $\tilde{f}(X)$.

The methods (b) and (c) are known interval arithmetic techniques. Applying them to the centred forms will always yield improvements of the centred forms although they are only significant if the domain X is not too small. We will now introduce these two techniques:

Interval arithmetic distinguishes between the simple and the extended power evaluation. Using the extended power evaluation, the symmetry of the powers of H is lost but the centred form is improved.

The *simple* version of the power evaluation (also called power evaluation by *simple arithmetic*) is defined by

$$A^0 = 1 \quad \text{and} \quad A^n = A \cdot \ldots \cdot A (n \text{ times}) \quad \text{if } n \geq 1.$$

The simple version of the power evaluation is the one that we have used up to this point. It will also be the version used in the sequel unless otherwise specified.

If A is symmetric, $A = [-a, a]$, then

$$A^n = [-a^n, a^n] \quad \text{if } n \geq 1,$$

that is, A^n is also symmetric.

The *extended* version of the power evaluation (also called power evaluation by *extended arithmetic*) is defined by

$$\tilde{A}^n = \{a^n : a \in A\} \quad \text{if } n \geq 0.$$

If A is symmetric, $A = [-a, a]$, then

$$\tilde{A}^n = \begin{cases} [-a^n, a^n] & \text{if } n \text{ is odd,} \\ [0, a^n] & \text{if } n \text{ is even, } n \neq 0. \end{cases}$$

The following two examples explain the use of two different kinds of power evaluation: If $x, y \in A$ then A^2 is the smallest interval that contains x, y. If $x \in A$ then $\tilde{A}^2$ is the smallest interval that contains x^2. Nevertheless, A^2 often is used for the inclusion of x^2 because the general assignment $A \to \tilde{A}^2$ is not a rational assignment, that is, it cannot be calculated using only the four interval arithmetic operations and constants.

If the powers that occur in an expression are evaluated using the extended version, then we shall indicate this by a tilde above this expression. This means that we write $\tilde{H}^n$, $\tilde{F}(X)$, $\tilde{Q}(X)$ if the extended arithmetic is used, and H^n, $F(X)$, $Q(X)$ if the simple arithmetic is used.

If $x \in X$, then we have $x - c \in X - c = H$ for $c = m(X)$ and by definition, $(x - c)^i \in \tilde{H}^i$ for $i = 0,1,\ldots,n$, such that $f(x) \in \bar{F}(X)$ and $\tilde{f}(X) \subseteq \bar{F}(X)$ follows if defined. This means that $\bar{F}(X)$ is also an including approximation of the range of f.

Since $\tilde{H}^i \subseteq H^i$ for $i = 0,1,\ldots,n$, the inclusion $\bar{F}(X) \subseteq F(X)$ follows by Lemma 1.4. This inclusion is nearly always a proper inclusion since $\tilde{H}^i \subseteq H^i$ is a proper inclusion when $i \neq 0$, i even. We recapitulate:

$$\tilde{H}^i \subseteq H^i \quad \text{for all non-negative } i, \tag{2.14}$$

$$\tilde{f}(X) \subseteq \bar{F}(X) \subseteq F(X). \tag{2.15}$$

Clearly, (2.14) and (2.15) are also valid if $c \neq m(X)$.

In order to construct interval functions which satisfy a Lipschitz condition it is important that the extended power evaluation

$$X \to \tilde{X}^n$$

is Lipschitz. This is a consequence of Theorem 1.3 since $X \to \tilde{X}^n$ is the range function of the function $f(x) = x^n$.

The *nested form* (*small Horner scheme*) is used in computational real analysis to evaluate polynomial values, since the influence of rounding errors is diminished. The nested form consists of a rearrangement of a polynomial in a certain manner, that is, if the polynomial or expression

$$p(x) = a_0 + a_1 x + \ldots + a_n x^n$$

is given, then the nested form of p is the expression

$$p_{\text{nest}}(x) = a_0 + x(a_1 + x(a_2 + x(a_3 + \ldots + x(a_{n-1} + x a_n) \ldots))).$$

It is clear that $p(x)$ and $p_{\text{nest}}(x)$ are identical as functions. The interval extensions of $p(x)$ and $p_{\text{nest}}(x)$ to an interval X satisfy the following inclusion:

$$p_{\text{nest}}(X) \subseteq p(X). \tag{2.16}$$

This inclusion is a direct consequence of the subdistributive law, (1.2).

Let us now apply (2.16) to the centred form while dropping the assumption $c = m(X)$. We arrange the expression $f(c) + s(x - c)$, see (2.6), such that the numerator and the denominator of this expression are developed in nested form. The natural interval extension of this new expression will be denoted by $F_{\text{nest}}(X)$ and is an including approximation of $\tilde{f}(X)$ by Theorem 1.1. The interval $F_{\text{nest}}(X)$ is furthermore contained in $F(X)$, by (2.16) and Lemma 1.4. We therefore have

$$\tilde{f}(X) \subseteq F_{\text{nest}}(X) \subseteq F(X). \tag{2.17}$$

If $c = m(X)$ as in the definition of the standard centred form it follows that

$$F_{nest}(X) = F(X),$$

that is, *the application of the nested form then only makes sense if $c \neq m(X)$*. The reason for this equality is that in (2.16), equality holds for symmetric X. This can be verified either by calculation using formulas (3) and (4) mentioned above, or as direct consequence of the validity of the distributive law in the case in question, see for example, Ratschek (1971).

We remark that it is not possible to use the nested form with extended arithmetic since each method needs its own specific rearrangement of F.

Example 2.4 Let again $f(x) = 1 - x + x^2$ and $X = [0, 2]$. We have seen from the Example 2.1 that

$$\tilde{f}(X) = [3/4, 3] \quad \text{and} \quad F(X) = [-1, 3].$$

Applying the extended arithmetic we get

$$\tilde{F}(X) = f(c) + t_1 H + t_2 \tilde{H}^2/2 = 1 + [-1, 1] + [0, 1] = [0, 3].$$

In order to demonstrate the workings of the nested form we choose $c = 0$. We may also construct the standard centred form using the parameter c and we obtain:

$$f(c) = 1, \quad t_1 = -1, \quad \text{and} \quad t_2 = 2,$$
$$H = X - c = [0, 2], \quad H^2 = \tilde{H}^2 = [0, 4].$$

Then

$$F(X) = \tilde{F}(X) = f(c) + t_1 H + t_2 H^2/2 = 1 + [-2, 0] + [0, 4]/2 = [-1, 3].$$

The nested form of the expresssion

$$s(x - c) = t_1(x - c) + t_2(x - c)^2/2 = -x + x^2$$

is $s_{nest}(x - c) = x(-1 + x)$ such that

$$F_{nest}(X) = 1 + X(-1 + X) = 1 + [0, 2][-1, 1] = [-1, 3].$$

We will now give a *plausibility argument* to show that, in the average, it is best to choose the midpoint of the domain X of f as developing point c in defining a centred form. We will not only discuss the standard centred form as defined above but we will also extend our consideration to each reasonable kind of a centred form of rational functions as defined later. Because of the rationality, each such centred form is an expression in the powers $(X - c)^r$ and some constants (coefficients). These coefficients depend on the function f, the domain X, the developing point c (which is

Sec. 2.3] Symmetric Intervals — or Not?

not necessarily the midpoint of X), and the current centred form. If we assume arbitrary (random) coefficients for f then the coefficients of the centred forms are also nearly arbitrary and random ones. We therefore focus on the powers $(X - c)^r$, these being the proper constituent parts of a centred form that are invariant over the class of all centred forms.

The following calculation shows that the smallest width of the powers $(X - c)^r$ is obtained (X is fixed) if c is the midpoint, and that the width increases with order r with respect to the difference $|c - m(X)|$.

We define the following quantities,

$c \in X$,
$X = m(X) + [-z, z]$,
$S = X - m(X) = [-z, z]$,
$d = c - m(X)$,
$H = X - c$.

Without loss of generality we assume that

$$m(X) \leq c \leq m(X) + z.$$

(The reason for choosing $c \in X$ is that this condition is needed for proving quadratic convergence, see Theorem 3.2.)

The last inequality is equivalent to

$$0 \leq d \leq z. \tag{2.18}$$

We will now compare the widths of the powers H^r and S^r if r is a natural number. They represent the powers occurring in the centred forms if any $c \in X$ or $m(X)$ is the developing point. Using the above assumptions we obtain

$S^r = [-z^r, z^r]$,
$w(S^r) = 2z^r$,
$H = X - c = X - m(X) + m(X) - c = S - d = [-z-d, z-d]$.

Using (2.18) the power of H is computed as

$$H^r = (-1)^r[d - z, d + z]^r = (-1)^r[(d - z)(d + z)^{r-1}, (d + z)^r].$$

Furthermore

$$w(H^r) = 2z(d + z)^{r-1}.$$

The difference of the widths is now

$$w(H^r) - w(S^r) = 2z[(d + z)^{r-1} - z^{r-1}]$$
$$= 2z \sum_{i=1}^{r-1} \binom{r-1}{i} d^i z^{r-1-i}$$
$$\geq \alpha(r - 1)z^{r-1}d \geq \beta d^r$$

for some constants α and β. The last estimation shows the dependency on d. The previous estimation shows the influence of the width.

2.4 STANDARD CENTRED FORMS OF HIGHER ORDER FOR FUNCTIONS IN ONE VARIABLE

In this section centred forms of higher order are defined. The name 'of higher order' originates from the use of Taylor expansions of order k in defining the standard centred form of order k, where k is any natural number larger than 0. The term 'kth order' does not imply that the convergence $w[F_k(X)] \to w[\bar{f}(X)]$ is of order k if $w(X) \to 0$, where the standard centred form of order k is denoted by $F_k(X)$. The convergence is again of order 2. The advantage of the form of order k is that a considerable improvement is obtained in comparison with the standard centred form defined earlier. The standard centred form turns out to be the form of first order. Furthermore, it is shown that the form of order k is better than the form of order $k - 1$. If the rational function is a polynomial, however, then the standard centred forms of all orders are equal such that in this case there is no proper improvement. The proofs are again only sketched since they are special cases of the proofs for functions in several variables. This distinction is made since the main ideas are clearer in the case of one variable due to the simpler technical machinery.

The background for defining new inclusions is again to rearrange the function $f(x) = f(c) + s(x - c)$ or $s(x - c)$ in a suitable manner and then to form the natural interval extension which is an inclusion of $\bar{f}(X)$ by Theorem 1.1.

Again, let $f = p/q$ be a rational function of one real variable represented by the quotient of the two polynomials p and q. Let n again be the maximum of the degrees of p and q. For a real compact interval X contained in the domain of f having the midpoint $c = m(X)$, and for any positive natural number k, we define for $\lambda = k,\ldots,k + n$ the coefficients,

$$t_{k\lambda} = p^{(\lambda)}(c) - \sum_{i=0}^{k-1} \binom{\lambda}{i} f^{(i)}(c) \, q^{(\lambda-i)}(c). \qquad (2.19)$$

These coefficients may also be written in the equivalent form,

$$t_{k\lambda} = \sum_{i=k}^{\lambda} \binom{\lambda}{i} f^{(i)}(c) \, q^{(\lambda-i)}(c). \qquad (2.20)$$

The equivalence of (2.19) and (2.20) is seen by substituting the λth derivative of $f(x)q(x)$ at c for $p^{(\lambda)}(c)$ in (2.19), that is using the formula

$$p^{(\lambda)}(c) = \sum_{i=0}^{\lambda} \binom{\lambda}{i} f^{(i)}(c) q^{(\lambda-i)}(c). \quad (2.21)$$

The coefficient $t_{k,k+n} = 0$ is not needed for defining centred forms. It is only used in comparisons like

$$t_{k\lambda} - t_{k-1,\lambda} = -\binom{\lambda}{k-1} f^{(k-1)}(c) q^{(\lambda-k+1)}(c) \quad (2.22)$$

which then is valid for every $k \geq 2$ and for each $\lambda = k,\ldots,k+n-1$. Let again $H = X - c = [-z, z]$.

Definition 2.2 The interval

$$F_k(X) = \frac{f(c) + f'(c)H + \ldots + f^{(k-1)}(c) H^{k-1}/(k-1)! + t_{kk}H^k/k! + \ldots + t_{k,k+n-1}H^{k+n-1}/(k+n-1)!}{q(c) + q'(c)H + \ldots + q^{(n)}(c)H^n/n!}$$

is called the *standard centred form of order k* for f on X if 0 does not lie in the denominator.

Clearly $F_k(X)$ is the natural interval extension of the right-hand side of the following expression:

$$f(x) = \sum_{\lambda=0}^{k-1} f^{(\lambda)}(c) h^\lambda/\lambda! + \left(\sum_{\lambda=k}^{k+n-1} t_{k\lambda}h^\lambda/\lambda!\right) \bigg/ \left(\sum_{\lambda=0}^{n} q^{(\lambda)}(c) h^\lambda/\lambda!\right) \quad (2.23)$$

where $h = x - c$. This means that in the interval extension x is replaced by X or h is replaced by H. Therefore we only have to verify the equality (2.23). Once that has been proven, the inclusions

$$\tilde{f}(X) \subseteq F_k(X) \quad \text{for } k = 1, 2, 3,\ldots \quad (2.24)$$

are valid by Theorem 1.1.

The formula (2.23) is nothing but the Taylor expansion of order k for f at c where the remainder is given by the quotient which replaces the usual representation of a remainder in terms of an unknown mean value $\xi \in X$. The verification of the equality (2.23) is accomplished by multiplying (2.23) by

$$q(x) = \sum_{\lambda=0}^{n} q^{(\lambda)}(c) h^\lambda/\lambda!,$$

and by substituting the terms (2.20) for the coefficients $t_{k\lambda}$ and performing some rearrangements. All the remaining factors can then be cancelled, without having to evaluate the derivatives $f^{(i)}(c)$.

Lemma 2.3 *The standard centred form $F_k(X)$ of order k for f on X is defined if and only if*

$$|q(c)| - |q'(c)|z - \ldots - |q^{(n)}(c)|z^n/n! > 0.$$

Proof The denominator is independent of k and it is equal to $Q(X)$ for all standard centred forms. The same conditions as given in Lemma 2.1 are therefore valid. □

Lemma 2.4 *The width of the standard centred form of order k is given by*

$$w[F_k(x)] = 2 \sum_{v=1}^{k-1} |f^{(v)}(c)| z^v/v! + 2 \left(\sum_{v=k}^{k+n-1} |t_{kv}| z^v/v! \right) \bigg/ \left(|q(c)| - \sum_{v=1}^{n} |q^{(v)}(c)| z^v/v! \right)$$

where $z = w(X)/2$.

Proof The lemma follows directly from the definition of the standard form of order k and from formulas (1.3) and (2.8). □

Remark 2.6 The standard centred form of f of first order is equal to the standard centred form.

Remark 2.7 If f is a polynomial, $f = p$, then $t_{kv} = p^{(v)}(c)$ and $Q(X) = 1$ such that

$$P(X) = P_k(X) \quad \text{for } k = 1, 2, \ldots .$$

Remarks 2.1 to 2.5 and also Section 2.3 are valid for standard centred forms of higher order. This means that the Taylor coefficients are again the constituent parts of $F_k(X)$. Moore's recursive technique may therefore again be used for their evaluation, see Remark 2.3. The computation of $F_k(X)$ is furthermore easily accomplished without access to interval software, see Remark 2.4. Again, it is possible to shift c from the midpoint. This is not recommended, see Remark 2.5 and Section 2.3. Extended arithmetic or, in the case of shifted c, the nested form, yields better inclusions, see Section 2.3.

In order to demonstrate that $F_k(X)$ is better than $F_{k-1}(X)$ we define for each natural number $k \geq 2$ and for $\lambda = k, k+1, \ldots, k+n-1$ the number $\varepsilon_{k\lambda}$ by

$$\varepsilon_{k\lambda} = |t_{k-1,\lambda}| - |t_{k\lambda}| + |t_{k-1,\lambda} - t_{k\lambda}|.$$

From the triangle equality written in the form

$$|t_{k\lambda}| = |t_{k\lambda} - t_{k-1,\lambda} + t_{k-1,\lambda}|$$
$$\leq |t_{k\lambda} - t_{k-1,\lambda}| + |t_{k-1,\lambda}|$$

it follows that

$$\varepsilon_{k\lambda} \geq 0.$$

Let $f = p/q$ be a rational function represented by the quotient of two polynomials p and q which have a maximum degree n. Furthermore, let X be a real compact interval lying in the domain of f and let $c = m(X)$. Then, provided the standard centred forms of all orders are defined (see Lemma 2.3), we have the following theorem:

Theorem 2.1 *The inequality*

$$w[F_k(X)] \leq w[F_{k-1}(X)]$$

is satisfied for $k = 2, 3, \ldots$.

Proof The inequality

$$0 \leq w[F_{k-1}(X)] - w[F_k(X)]$$

is rearranged in such a manner that its validity is immediately obvious. First, the widths are replaced by the formulas given in Lemma 2.4. Then, the inequality is multiplied by the denominator. This does not change the direction of the inequality since the denominator is positive from Lemma 2.3. The factor 2 is now cancelled from the inequality, leaving

$$0 \leq \sum_{\lambda=k}^{k+n-1} \varepsilon_{k\lambda} z^\lambda/\lambda! \quad \text{if } z = w(X)/2$$

which is valid since $\varepsilon_{k\lambda} \geq 0$ and $z \geq 0$. $\square$

Corollary 2.1 *For each* $k = 2, 3, \ldots$ *the inclusions* $\tilde{f}(X) \subseteq F_k(X) \subseteq F_{k-1}(X)$ *are valid.*

Proof The first inclusion is valid because of (2.24), the second one follows from the representation (2.11) and Theorem 2.1. $\square$

Theorem 2.1 and Corollary 2.1 do not express whether the improvements gained by the forms of higher order are proper improvements or not. A large number of examples which cannot be reproduced here as well as the following consideration demonstrate that these improvements are in general proper:

Theorem 2.2 *Let* $k \geq 2$ *be a natural number and* $z = w(X)/2 \neq 0$. *Then the equality* $F_k(X) = F_{k-1}(X)$ *holds if and only if*

$$\varepsilon_{k\lambda} = 0 \text{ for } \lambda = k, \ldots, k + n - 1.$$

Proof Since $f(c)$ is the midpoint of all forms of higher order the equality $F_k(X) = F_{k-1}(X)$ is equivalent to the equation $w[F_k(X)] = w[F_{k-1}(X)]$

which is furthermore equivalent to

$$0 = \sum_{\lambda=k}^{k+n-1} \varepsilon_{k\lambda} z^\lambda / \lambda! \qquad (2.25)$$

adopting the rearrangement described in the proof of Theorem 2.1. Because of the assumption $z \neq 0$ equation (2.25) holds if and only if all the coefficients $\varepsilon_{k\lambda} = 0$. □

Remark 2.8 Theorem 2.2 shows that the inclusion $F_k(X) \subseteq F_{k-1}(X)$ is in general proper because the construction of the numbers $\varepsilon_{k\lambda}$ is such that the condition of Theorem 2.2 is usually not satisfied. The improvements gained by the centred forms of higher order are therefore significant. There is one systematic exception: If f is a polynomial then there is no improvement (see Remark 2.7). In this case the reader may wish to verify that

$$\varepsilon_{k\lambda} = 0 \quad \text{for } \lambda = k,\ldots,k+n-1.$$

Remark 2.9 A partially recursive computation of the standard forms of higher order is possible. In order to proceed from $F_{k-1}(X)$ to $F_k(X)$ it is not necessary to compute the coefficients $t_{k\lambda}$ from scratch. One may use the already known coefficients $t_{k-1,\lambda}$ in order to compute $t_{k\lambda}$ using (2.22). A more refined technique for recursive computation of these forms will be described in Section 2.5 where the standard forms for functions in several variables are treated.

Remark 2.10 It is interesting to note that the inclusion chain $F_k(X) \subseteq F_{k-1}(X)$ is no longer valid if the extended power evaluation is used, see Corollary 2.1. The following counter-example shows that the conjectures $\tilde{F}_k(X) \subseteq \tilde{F}_{k-1}(X)$ as well as the conjectures $\tilde{F}_k(X) \subseteq \tilde{F}_{k-2}(X)$ are false.

Example 2.4 Let $f(x) = (1 + x + x^2)/(1 + x)$ be defined on the domain $X = [1,3]$. Then we get the following values:

$w[\tilde{f}(X)] = 1.75$

k	$w[\tilde{F}_k(X)]$
1	3.16̇6̇
2	1.83̇3̇
3	1.85 1̇8̇5̇
4	1.84567 9012̇

(All computed results are exact; 3̇, 6̇, and 1̇85̇, etc. indicate periodic decimals.)

Sec. 2.4]   **Higher Order for Functions in One Variable**   47

Remark 2.11 From Definition 2.2 and from Corollary 2.1 it follows that $\lim_{k \to \infty} F_k(X)$ exists if $F_1(X)$ exists. In this case the limes has the value

$$\sum_{k=0}^{\infty} f^{(k)}(c) H^k/k!.$$

One can check without difficulty that $\tilde{f}(X) = \lim_{k \to \infty} F_k(X)$ iff $f^{(k)}(c) \geq 0$ for all $k \geq 1$.

In the following examples we show the result of the computation of the standard centred forms of higher order for domains having small and large widths.

We note that the given results of the following examples were first computed using a machine interval arithmetic (see, for example, Apostolatos–Kulisch, 1967) and a 28-bit machine number representation with an additional exponent part. These intermediate results were copies from the computer print-outs to a 4-digit representation. The printed interval results are therefore falsified by rounding errors but they include in any case the exact interval result. The printed results for the widths are rounded up in the normal manner.

Example 2.5 The function

$$f(x) = (3x^4 + 3x^3 + x^2 + 3x + 3)/(3x^3 + x^2 + 4x + 2)$$

is given. We first compute inclusions of the range of this function over $X = [0.9, 1.1]$ with the help of the standard centred forms:

k	$F_k(X)$	$w[F_k(X)]$
1	[1.203, 1.397]	0.1933
2	[1.228, 1.372]	0.1437
3	[1.229, 1.371]	0.1414
4	[1.229, 1.371]	0.1412

We furthermore calculate

$$\tilde{f}(X) \simeq [1.240, 1.370], \quad w[\tilde{f}(X)] \simeq 0.1294$$

$$f(X) = [0.9191, 1.849], \quad w[f(X)] = 0.9298$$

(see remark in Example 2.2 on the approximation of $\tilde{f}(X)$). We now compute the standard centred forms over [0.9, 1.1] using the extended interval arithmetic and we obtain

Standard Centred Form [Ch. 2]

k	$\tilde{F}_k(X)$	$w[\tilde{F}_k(X)]$
1	[1.222, 1.396]	0.1733
2	[1.234, 0.372]	0.1373
3	[1.234, 1.371]	0.1361

The range of the same function is now included over the interval $X = [0.6, 1.4]$ using the standard centred forms and first simple and then extended interval arithmetic. The results are given below.

k	$F_k(X)$	$w[F_k(X)]$	$\tilde{F}_k(X)$	$w[\tilde{F}_k(X)]$
1	[−1.322, 3.922]	5.243	[0.4306, 2.820]	2.390
2	[0.5140, 2.086]	1.572	[0.9626, 1.865]	0.9023
3	[0.8050, 1.795]	0.9899	[0.9524, 1.704]	0.7508
4	[0.9170, 1.683]	0.7659	[1.014, 1.676]	0.6610
5	[0.9296, 1.671]	0.7408	[1.017, 1.669]	0.6517
6	[0.9306, 1.670]	0.7388	[1.017, 1.668]	0.6506
7	[0.9319, 1.6681]	0.7361	[1.018, 1.668]	0.6496
8	[0.932188, 1.66782]	0.735625	[1.01808, 1.66751]	0.649418
9	[0.932195, 1.66781]	0.735609	[1,01806, 1.66750]	0.649440
10	[0.93226, 1.66778]	0.735550	[1.01808, 1.66750]	0.649415

The approximation to the range (see Example 2.2) is $\tilde{f}([0.6, 1.4]) \simeq [1.145, 1.626]$ such that $w[\tilde{f}([0.6, 1.4])] = 0.4795$. Furthermore, $f([0.6, 1.4]) = [0.3482, 5.348]$ with $w[f([0.6, 1.4])] = 4.999$.

From the above results as well as further experimentation we note that if the domain is wide then it makes sense to compute the higher order ($k = 3$, 4, 5) centred forms as significant improvements may be noticed. For small widths of the domain it is probably better to compute only the second order form.

Example 2.6 Let

$$f(x) = (x^5 + x^4 + x^3 - x^2 - x + 1)/(x^4 + 2x^3 + x^2 + 2x + 1)$$

be a given rational function and let $X_1 = [0.03, 0.1]$ and $X_2 = [0.15, 0.5]$ such that $w(X_1) = 0.07$ and $w(X_2) = 0.35 = 5w(X_1)$. The following results were obtained.

Sec. 2.4] Higher Order for Functions in One Variable 49

	$i=1$	$i=2$
$w[\bar{f}(X_i)]$	0.1783	0.4424
$w[f(X_i)]$	0.1790	0.5224
$w[F_1(X_i)]$	0.1956	0.7641
$w[F_2(X_i)]$	0.1865	0.7358
$w[F_3(X_i)]$	0.1860	0.5728
$w[F_4(X_i)]$	0.1860	0.5769
$w[F_5(X_i)]$	0.1860	0.5769
$w[F_6(X_i)]$	0.1860	0.5768
$w[\bar{F}_1(X_i)]$	0.1930	0.6865
$w[\bar{F}_2(X_i)]$	0.1824	0.6028
$w[\bar{F}_3(X_i)]$	0.1822	0.5136
$w[\bar{F}_4(X_i)]$	0.1822	0.5098
$w[\bar{F}_5(X_i)]$	0.1822	0.5098
$w[\bar{F}_6(X_i)]$	0.1822	0.5096

In the above table we note that the natural interval extension $f(X)$ is better than the centred forms $F_k(X)$, $k = 1,\ldots,6$.

2.5 STANDARD CENTRED FORMS FOR FUNCTIONS IN SEVERAL VARIABLES

The ideas of the last section are now carried over to functions of several variables. Since the motivations are the same as for the centred forms in one variable we eschew these here in order to concentrate on the technical developments.

Let f be a rational function of m real variables $x = (x_1,\ldots,x_m)$ and let $X = X_1 \times \ldots \times X_m$ be any right parallelepiped lying in the domain of f where $X_i \in I$ for $i = 1,\ldots,m$. Let the function f again be represented as quotient of the polynomials p and q,

$$f = p/q,$$

and let n denote the maximum degree of p and q.

Explicit rational expressions $F_k(x)$ are given which define the function f. Their natural interval extensions $F_k(X)$ are called standard centred forms of order k of f. It is again shown that $F_k(X) \supseteq \bar{f}(X)$ and that $F_k(X)$ is an improvement over $F_{k-1}(X)$.

We employ multi-indices in order to avoid involved formulas. Multi-indices are vectors $\lambda, \rho, \mu \in N^m$, where $N = \{0, 1, 2,\ldots\}$. The notations $\lambda = (\lambda_1,\ldots,\lambda_m)$ and $o = (0,\ldots,0)$ are used. The following common abbreviations are also used here:

$$|\lambda| = \lambda_1 + \ldots + \lambda_m, \quad \lambda! = \lambda_1! \cdot \ldots \cdot \lambda_m!,$$

$$A^\lambda = A_1^{\lambda_1} A_2^{\lambda_2} \cdot \ldots \cdot A_m^{\lambda_m} \quad \text{for } A \in R^m \text{ or } A \in I^m \text{ or } A = x \text{ where } A_i \text{ are the components of } A,$$

$$\binom{\lambda}{\rho} = \binom{\lambda_1}{\rho_1}\binom{\lambda_2}{\rho_2} \cdot \ldots \cdot \binom{\lambda_m}{\rho_m},$$

$$D^\lambda f(x) = \frac{\partial^{\lambda_1 + \ldots + \lambda_m} f(x)}{\partial x_1^{\lambda_1} \cdot \ldots \cdot \partial x_m^{\lambda_m}}$$

Furthermore, $\lambda \leq \rho$ denotes $\lambda_i \leq \rho_i$, for $i = 1,2,\ldots m$. The symbol $\sum_{|\lambda|=0}^{k} b_\lambda$ for some $k \in N$ denotes the sum over all b_λ with $|\lambda| = 0,1,\ldots,k$. Similarly, the notation $\sum_{\lambda=o}^{\rho} b_\lambda$ denotes the sum over all b_λ where λ satisfies $o \leq \lambda \leq \rho$, etc.

We need some further abbreviations. Let c_i be the midpoint of the interval X_i and $c = (c_1,\ldots,c_m)$. If $X_i = [c_i - z_i, c_i + z_i]$ for some non-negative real z_i, then $H_i = X_i - c_i = [-z_i, z_i]$ is symmetric for each i. The interval $H = X - c = [-z, z]$ is now used in the calculations.

For $\lambda \in N^m$ and any positive integer $k \leq |\lambda|$ we define

$$t_{k\lambda} = D^\lambda p(c) - \sum_{|\rho|=0, \rho \leq \lambda}^{k-1} \binom{\lambda}{\rho} D^\rho f(c) D^{\lambda-\rho} q(c) \tag{2.26}$$

(or equivalently,

$$t_{k\lambda} = \sum_{|\nu|=k, \rho \leq \lambda}^{|\lambda|} \binom{\lambda}{\rho} D^\rho f(c) D^{\lambda-\rho} q(c)). \tag{2.27}$$

The equivalent formula arises directly if $D^\lambda p(c)$ is replaced by $D^\lambda[f(c)q(c)]$:

$$D^\lambda p(c) = \sum_{\rho=o}^{\lambda} \binom{\lambda}{\rho} D^\rho f(c) D^{\lambda-\rho} q(c). \tag{2.28}$$

The formula (2.28) is nothing more than the product rule for higher partial derivatives.

Definition 2.3 Let k be a positive integer. Then the interval

$$F_k(X) = f(c) + \sum_{|\lambda|=1}^{k-1} D^\lambda f(c) H^\lambda/\lambda! + \frac{\sum_{|\lambda|=k}^{k+n-1} t_{k\lambda} H^\lambda/\lambda!}{\sum_{|\lambda|=0}^{n} D^\lambda q(c) H^\lambda/\lambda!}$$

is called the *standard centred form of order k* for the function f on X if 0 does not lie in the denominator.

Sec. 2.5] Standard Centred Forms for Functions in Several Variables 51

For instance, the standard centred form (of first order) as defined by Ratschek (1978) is the interval

$$F(X) = F_1(X) = f(c) + \frac{\sum_{|\lambda|=1}^{n} t_{1\lambda} H^\lambda/\lambda!}{\sum_{|\lambda|=0}^{n} D^\lambda q(c) H^\lambda/\lambda!}$$

In order to show that $F_k(X)$ is an outer approximation of $\tilde{f}(X)$ we first notice that $F_k(X)$ is the natural interval extension of the rational expression

$$F_k(x) = f(c) + \sum_{|\lambda|=1}^{k-1} D^\lambda f(c) \frac{(x-c)^\lambda}{\lambda!} + \frac{\sum_{|\lambda|=k}^{k+n-1} t_{k\lambda}(x-c)^\lambda/\lambda!}{\sum_{|\lambda|=0}^{n} D^\lambda q(c)(x-c)^\lambda/\lambda!}. \qquad (2.29)$$

This means that we only have to prove that F_k and f are identical.

Lemma 2.5 *The functions f and F_k are identical.*

Proof Write h instead of $x - c$ for compactness and define

$$L(h) = \sum_{|\lambda|=0}^{n} D^\lambda q(c) h^\lambda/\lambda!,$$

$$M(h) = \sum_{|\lambda|=1}^{n} [D^\lambda p(c) - f(c) D^\lambda q(c)] h^\lambda/\lambda!.$$

The function $L(h)$ is the Taylor expansion of q at the point c. Applying the Taylor's formula of order k to f and denoting the remainder by $g(h)$, we obtain

$$f(c+h) - f(c) = \sum_{|\lambda|=1}^{k-1} D^\lambda f(c) h^\lambda/\lambda! + g(h). \qquad (2.30)$$

In order to obtain the exact value of $g(h)$ the value of $f(c+h) - f(c)$ is calculated in a different manner, by writing the terms $p(c+h)$ and $q(c+h)$ as Taylor expansions:

$$f(c+h) - f(c) = \frac{p(c+h) - f(c)q(c+h)}{q(c+h)} = \frac{M(h)}{L(h)}. \qquad (2.31)$$

In identifying (2.30) and (2.31) we get an explicit formula for $g(h)$, namely the quotient of the two sums in (2.29),

$$g(h) = \frac{M(h)}{L(h)} - \sum_{|\lambda|=1}^{k-1} D^\lambda f(c) h^\lambda/\lambda!.$$

The details of the proof are now as follows. First, the following term is rearranged:

$$L(h) \sum_{|\lambda|=1}^{k-1} D^\lambda f(c) \frac{h^\lambda}{\lambda!} \tag{2.32}$$

$$= \left(\sum_{|\mu|=1}^{k-1} D^\mu f(c) \frac{h^\mu}{\mu!} \right) \left(\sum_{|\lambda|=0}^{n} D^\lambda q(c) \frac{h^\lambda}{\lambda!} \right)$$

$$= \sum_{|\mu|=1}^{k-1} \sum_{|\lambda|=0}^{n} D^\mu f(c) D^\lambda q(c) \frac{h^{\mu+\lambda}}{\mu!\lambda!}$$

$$= \sum_{|\mu|=1}^{k-1} \sum_{\substack{|\nu|=|\mu| \\ \nu \geq \mu}}^{|\mu|+n} D^\mu f(c) D^{\nu-\mu} q(c) \frac{h^\nu}{\mu!(\nu-\mu)!}$$

$$= \sum_{|\mu|=1}^{k-1} \sum_{\substack{|\nu|=|\mu| \\ \mu \leq \nu}}^{k-1+n} D^\mu f(c) D^{\nu-\mu} q(c) \binom{\nu}{\mu} \frac{h^\nu}{\nu!}$$

$$= \sum_{|\nu|=1}^{k-1+n} \sum_{\substack{|\mu|=1 \\ \nu \geq \mu}}^{k-1} D^\mu f(c) D^{\nu-\mu} q(c) \binom{\nu}{\mu} \frac{h^\nu}{\nu!}$$

$$= \sum_{|\nu|=1}^{k-1} \sum_{\substack{|\mu|=1 \\ \mu \leq \nu}}^{k-1} D^\mu f(c) D^{\nu-\mu} q(c) \binom{\nu}{\mu} \frac{h^\nu}{\nu!} +$$

$$\sum_{|\nu|=k}^{k-1+n} \sum_{\substack{|\mu|=1 \\ \mu \leq \nu}}^{k-1} D^\mu f(c) D^{\nu-\mu} q(c) \binom{\nu}{\mu} \frac{h^\nu}{\nu!}$$

$$= \sum_{|\lambda|=1}^{k-1} \sum_{\substack{|\rho|=1 \\ \rho \leq \lambda}}^{k-1} D^\rho f(c) D^{\lambda-\rho} q(c) \binom{\lambda}{\rho} \frac{h^\lambda}{\lambda!} +$$

$$\sum_{|\nu|=k}^{k-1+n} \sum_{\substack{|\mu|=1 \\ \mu \leq \nu}}^{k-1} D^\mu f(c) D^{\nu-\mu} q(c) \binom{\nu}{\mu} \frac{h^\nu}{\nu!}$$

Using (2.28) and considering the equation

$$D^\lambda p(c) = 0 \quad \text{for } |\lambda| > n$$

the expression for $M(h)$ is rearranged:

$$M(h) = \sum_{|\lambda|=1}^{n} \left[\sum_{\rho=0}^{\lambda} D^\rho f(c) D^{\lambda-\rho} q(c) \binom{\lambda}{\rho} - f(c) D^\lambda q(c) \right] \frac{h^\lambda}{\lambda!}$$

$$= \sum_{|\lambda|=1}^{n} \left[\sum_{\substack{\rho=0 \\ |\rho| \geq 1}}^{\lambda} D^\rho f(c) D^{\lambda-\rho} q(c) \binom{\lambda}{\rho} \right] \frac{h^\lambda}{\lambda!}$$

$$= \sum_{|\lambda|=1}^{k-1} \left[\sum_{\substack{\rho=0 \\ |\rho| \geq 1}}^{\lambda} D^\rho f(c) D^{\lambda-\rho} q(c) \binom{\lambda}{\rho} \right] \frac{h^\lambda}{\lambda!} +$$

$$\sum_{|\lambda|=k}^{n} \left[\sum_{\substack{\rho=0 \\ |\rho|=1}}^{\lambda} D^\rho f(c) D^{\lambda-\rho} q(c) \binom{\lambda}{\rho} \right] \frac{h^\lambda}{\lambda!}$$

Sec. 2.5] Standard Centred Forms for Functions in Several Variables 53

Using this expression for $M(h)$ and remembering (2,32), we obtain:

$$\begin{aligned}
g(h) &= \frac{1}{L(h)}\left[M(h) - L(h)\sum_{|\lambda|=1}^{k-1}D^\lambda f(c)\frac{h^\lambda}{\lambda!}\right]\\
&= \frac{1}{L(h)}\left\{\sum_{|\lambda|=k}^{n}\left[\sum_{\substack{\rho=o\\|\rho|\geq 1}}^{\lambda}D^\rho f(c)D^{\lambda-\rho}q(c)\binom{\lambda}{\rho}\right]\frac{h^\lambda}{\lambda!}\right.\\
&\quad -\sum_{|\nu|=k}^{n}\sum_{\substack{|\mu|=1\\ \mu\leq\nu}}^{k-1}D^\mu f(c)D^{\nu-\mu}q(c)\binom{\nu}{\mu}\right]\frac{h^\nu}{\nu!}\\
&\quad \left.-\sum_{|\nu|=n+1}^{k-1+n}\left[\sum_{\substack{|\mu|=1\\ \mu\leq\nu}}^{k-1}D^\mu f(c)D^{\nu-\mu}q(c)\binom{\nu}{\mu}\right]\frac{h^\nu}{\nu!}\right\}\\
&= \frac{1}{L(h)}\left[\sum_{|\lambda|=k}^{n}\sum_{\substack{|\rho|=k\\ \rho\leq\lambda}}^{|\lambda|}\binom{\lambda}{\rho}D^\rho f(c)D^{\lambda-\rho}q(c)\frac{h^\lambda}{\lambda!}\right.\\
&\quad -\sum_{|\nu|=n+1}^{k-1+n}\sum_{\substack{|\mu|=1\\ \mu\leq\nu}}^{k-1}\binom{\nu}{\mu}D^\mu f(c)D^{\nu-\mu}q(c)\frac{h^\nu}{\nu!}\\
&= \frac{1}{L(h)}\left[\sum_{|\lambda|=k}^{n}\sum_{\substack{|\rho|=k\\ \rho\leq\lambda}}^{|\lambda|}D^\rho f(c)D^{\lambda-\rho}q(c)\binom{\lambda}{\rho}\frac{h^\lambda}{\lambda!}\right.\\
&\quad \left.+\sum_{|\lambda|=n+1}^{k-1+n}\sum_{\substack{|\rho|\geq k\\ \rho\leq\lambda}}^{|\lambda|}D^\rho f(c)D^{\lambda-\rho}q(c)\binom{\lambda}{\rho}\frac{h^\lambda}{\lambda!}\right]\\
&= \frac{1}{L(h)}\sum_{|\lambda|=k}^{k-1+n}\sum_{\substack{|\rho|=k\\ \rho\leq\lambda}}^{|\lambda|}\binom{\lambda}{\rho}D^\rho f(c)D^{\lambda-\rho}q(c)\frac{h^\lambda}{\lambda!}
\end{aligned}$$

Inserting this term into (2.30) gives the expression $F_k(x)$. □

Theorem 2.3 *For all positive integers k the inclusion $\tilde{f}(X) \subseteq F_k(X)$ holds.*

Proof This assertion is an immediate consequence of Lemma 2.5 and Theorem 1.1. □

Lemma 2.6 (*Ratschek–Schröder, 1981*). *For every positive integer k, the standard centred form of order k is defined if and only if*

$$|q(c)| - \sum_{|\lambda|=1}^{n}|D^\lambda q(c)|z^\lambda/\lambda! > 0.$$

Proof The existence of the form of order k is independent on k because all forms have the same denominator, $L(z)$, as defined in the proof of

Lemma 2.5. The condition

$$0 \notin L(z)$$

is equivalent to the above inequality because of (2.7). □

Lemma 2.7 (Ratschek-Schröder, 1981). *The width of the standard centred form of order k is given by*

$$w[F_k(X)] = 2 \sum_{|\lambda|=1}^{k-1} |D^\lambda f(c)| \frac{z^\lambda}{\lambda!} + 2 \frac{\sum_{|\lambda|=k}^{k+n-1} |t_{k\lambda}| z^\lambda/\lambda!}{|q(c)| - \sum_{|\lambda|=1}^{n} |D^\lambda q(c)| z^\lambda/\lambda!}$$

Proof The assertion is a direct consequence of the definition of a form of order k and of formulas (1.3) and (2.8). □

Theorem 2.4 *For each natural number $k \geq 2$ the inequality*

$$w[F_k(X)] \leq w[F_{k-1}(X)]$$

is satisfied.

Proof There is a simple arithmetic relation between the coefficients of order $k-1$ and order k, given by:

$$t_{k\lambda} - t_{k-1,\lambda} = - \sum_{\substack{|\rho|=k \\ \rho \leq \lambda}} D^\rho f(c) D^{\lambda-\rho} q(c) \binom{\lambda}{\rho}. \quad (2.33)$$

This follows directly from (2.26). We now define

$$K = \sum_{|\lambda|=1}^{n} |D^\lambda q(c)| z^\lambda!$$

where $z = w(X)/2$. Since $|q(c)| > K$ from Lemma 2.6, it suffices to show that

$$d = \{w[F_{k-1}(X)] - w[F_k(X)]\}[|q(c)| - K]/2 \geq 0. \quad (2.34)$$

First, we perform the following calculation in which the notation is changed twice. The first time μ is substituted for λ's. The second change in notation is that μ is replaced by $\lambda - \rho$:

$$\left(\sum_{|\lambda|=1}^{n} |D^\lambda q(c)| z^\lambda/\lambda! \right) \sum_{|\rho|=k-1} |D^\rho f(c)| z^\rho/\rho!$$

$$= \sum_{|\rho|=k-1} \sum_{|\mu|=1}^{n} |D^\rho f(c) D^\mu q(c)| z^{\rho+\mu}/(\rho!\mu!)$$

$$= \sum_{|\rho|=k-1}^{k+n-1} \sum_{\substack{|\lambda|=k \\ \lambda \geq \rho}}^{k+n-1} |D^\rho f(c) D^{\lambda-\rho} q(c)| \frac{\lambda!}{\rho!(\lambda-\rho)!} \frac{z^\lambda}{\lambda!}$$

$$= \sum_{|\lambda|=k}^{k+n-1} \sum_{\substack{|\rho|=k-1 \\ \rho \leq \lambda}} |D^\rho f(c) D^{\lambda-\rho} q(c)| \binom{\lambda}{\rho} \frac{z^\lambda}{\lambda!}$$

$$\geq \sum_{|\lambda|=k} \left| \sum_{\rho \leq \lambda} D^\rho f(c) D^{\lambda-\rho} q(c) \binom{\lambda}{\rho} \right| \frac{z^\lambda}{\lambda!}$$

$$= \sum_{|\lambda|=k} |t_{k-1,\lambda} - t_{k,\lambda}| \frac{z^\lambda}{\lambda!} \qquad (2.35)$$

With the help of the estimation for (2.35) developed above, the desired inequality (2.34) can be derived:

$$d = -\sum_{|\lambda|=k-1} |D^\lambda f(c) q(c)| z^\lambda/\lambda! + K \sum_{|\lambda|=k-1} |D^\lambda f(c)| z^\lambda/\lambda!$$

$$+ \sum_{|\lambda|=k-1} |t_{k-1,\lambda}| z^\lambda/\lambda! + \sum_{|\lambda|=k}^{k+n-1} (|t_{k-1,\lambda}| - |t_{k\lambda}|) z^\lambda/\lambda!$$

$$\geq \sum_{|\lambda|=k}^{k+n-1} |t_{k-1,\lambda} - t_{k\lambda}| \frac{z^\lambda}{\lambda!} + \sum_{|\lambda|=k}^{k+n-1} (|t_{k-1,\lambda}| - |t_{k\lambda}|) \frac{z^\lambda}{\lambda!}$$

$$\geq 0.$$

The last inequality follows from the triangle inequality. □

Corollary 2.2 *For each $k = 2, 3, \ldots$ the inclusions $\bar{f}(X) \leq F_k(X) \leq F_{k-1}(X)$ are valid.*

Proof The first inclusion is just Theorem 2.3 and the second inclusion follows from the representation (2.11) and Theorem 2.4. □

Remark 2.12 If f is a polynomial, $f = p$, then $t_{k\lambda} = D^\lambda p(c)$ and $Q(X) = 1$ such that

$$P(X) = P_k(X) \qquad \text{for } k = 1, 2, \ldots .$$

The proof of Theorem 2.4 shows, however, that the inequality $w[F_k(X)] \leq w[F_{k-1}(X)]$ is in general a proper inequality. The improvements obtained by the use of forms of higher order are therefore significant.

Remark 2.13 If $F_k(X)$ exists for some k, then $\lim_{k \to \infty} F_k(X)$ exists and has the value

$$\sum_{|\lambda|=0}^{\infty} D^\lambda f(c) H^\lambda/\lambda!,$$

as in the case of one variable. The equality $\lim_{k \to \infty} F_k(X) = \bar{f}(X)$ holds iff $D^\lambda f(c) \geq 0$ for all $|\lambda| \geq 1$.

Remark 2.14 It is possible to compute $F_k(X)$ in a recursive manner. This computation consists of three parts:

(a) The constituent parts of the coefficients $t_{k\lambda}$ are again the Taylor coefficients. We can therefore use Moore's technique for the recursive computation of Taylor coefficients without writing the coefficients down or programming them explicitly, see Moore (1966, 1979). Since we use multiindices the formulas for functions in several variables are almost the same as the formulas for one variable as described in Moore (1979). Explicit numerical data for the complexity of the computation of Taylor coefficients can easily be derived from the concept given in Moore (1979, pages 24 and 40), see also Rall (1981).

(b) In addition to (a), the coefficients $t_{k\lambda}$ can be calculated recursively using formula (2.33).

(c) The recursive computation of $F_k(X)$ is possible in the following manner: In order to proceed from $F_{k-1}(X)$ to $F_k(X)$, we write $F_{k-1}(X)$ as

$$F_{k-1}(X) = S + \left(T + \sum_{|\lambda|=k}^{k+n-2} t_{k-1,\lambda} H^\lambda/\lambda!\right)/U,$$

where

$$S = f(c) + \sum_{|\lambda|=1}^{k-2} D^\lambda f(c) H^\lambda/\lambda!,$$

$$T = \sum_{|\lambda|=k-1} t_{k-1,\lambda} H^\lambda/\lambda! = q(c) \sum_{|\lambda|=k-1} D^\lambda f(c) H^\lambda/\lambda!,$$

$$U = \sum_{|\lambda|=0}^{n} D^\lambda q(c) H^\lambda/\lambda!.$$

The values S, T, U, $t_{k-1,\lambda}$, and $H^\lambda/\lambda!$ should be recorded. Using the definition of $F_k(X)$ it follows that

$$F_k(X) = S + (T/q(c) + \sum_{|\lambda|=k-1}^{k+n-1} t_{k\lambda} H^\lambda/\lambda!)/U,$$

$$t_{k\lambda} = t_{k-1,\lambda} - \sum_{\substack{|\rho|=k-1 \\ \rho \leq \lambda}} D^\rho f(c) D^{\lambda-\rho} q(c) \binom{\lambda}{\rho},$$

which means that the computational effort is relatively small. The new values $D^\rho f(c)$ occurring for $|\rho| = k - 1$ may also be computed recursively, see (a) above. The binomial coefficients, factorials, and, if simple arithmetic is used, the powers H^λ may similarly be computed.

If the evaluation of $F_k(X)$ is reduced to the form of order $k - 2$ or lower then the same principles as described in (a) are applicable, the only change being that the splitting of the sums must be finer.

We note that all the remarks of the former sections as well as the content of Section 2.3 are also valid for standard centred forms for functions in several variables. It should be remembered that Lemmas 2.6 and 2.7, Theorem 2.4, and Corollary 2.2 are only proven if $c = m(X)$. One additional remark is to mention a problem which arises mainly in case of several variables:

If we are given an expression for a rational function $f(x)$ then there does not seem to exist a general criterion that states whether the interval

Sec. 2.5] Standard Centred Forms for Functions in Several Variables 57

extension $f(X)$ of this function provides a better or worse inclusion for $\bar{f}(X)$ than $F_k(X)$. Some results are, however, known as to when the interval extension $f(X)$ gives the exact range, $\bar{f}(X)$, and is therefore equal to or better than $F_k(X)$. This is for instance the case, if in the rational expression $f(x)$ each variable occurs only once and only to the first power, see Moore (1979). More generally, if some of the variables $x_1,\ldots,x_m$ occur only once in the expression $f(x)$, then these variables need not be taken into account when developing the centred form, see Skelboe (1974) and Moore (1976). These points are discussed extensively in Chapter IV.

Example 2.7 Some numerical examples are now given. Let $f(x_1, x_2) = 1/(x_1^2 + x_1 + x_2^2 + x_2 + 1)$, $X = [-0.1, 0.1] \times [-0.1, 0.1]$, and $Y = [-0.01, 0.01] \times [-0.01, 0.01]$. Then we obtain the following values:

$$w[\bar{f}(X)] = 0.3\ 9983\ 90884, \quad w[\bar{f}(Y)] = 0.03\ 9999\ 99841$$

k	$w(\tilde{F}_k(X))$	$w[\tilde{F}_k(Y)]$
1	0.5 2500 00000	0.04 1020 40816
2	0.4 6000 00000	0.04 0416 32653
3	0.4 5100 00000	0.04 0408 24489
4	0.4 4960 00000	0.04 0408 12326
5	0.4 4942 50000	0.04 0408 12165
6	0.4 4939 95000	0.04 0408 12163

Example 2.8 Let the function

$$f(x_1, x_2) = (1 + x_1^2 x_2)/(2 - x_1 x_2)$$

be defined on the interval

$$X_i = \left[-\frac{1}{i}, \frac{1}{i}\right] \times \left[-\frac{1}{i}, \frac{1}{i}\right] \quad \text{for } i = 1, 2, 4, 10, | 100, 1000.$$

The ranges are:

i	$\bar{f}(X_i)$	$w[\bar{f}(X_i)]$
1	[0, 2]	2
2	[0.38, 0.6428571]	0.2539685397
4	[0.4772, 0.5241935484]	0.0469208212
10	[0.4970149253, 0.5030150754]	0.0060001501
100	[0.4999745012, 0.5000255013]	0.0000510001
1000	[0.4999997495, 0.5000002506]	0.0000005011

The following numerical results for $\tilde{F}_k(X_i)$ and $w[\tilde{F}_k(X_i)]$ were obtained by Rausch (1981).

i	$\tilde{F}_1(X_i) = \tilde{F}_2(X_i)$	$w[\tilde{F}_1(X_i)]$
1	$[-1.0, 2.0]$	3.0
2	$[0.3571428751, 0.6428571429]$	0.2857142858
4	$[0.4758064516, 0.5241935484]$	0.0483870968
10	$[0.4969849246, 0.5030150754]$	0.0060301508
100	$[0.4999744987, 0.5000255013]$	0.0000510025
1000	$[0.4999997495, 0.5000002506]$	0.0000005011

i	$\tilde{F}_3(X_i)$	$w[\tilde{F}_3(X_i)]$
1	$[-0.75, 2.0]$	2.75
2	$[0.3660714286, 0.6428571429]$	0.2767857143
4	$[0.4763104839, 0.5241935484]$	0.0478830645
10	$[0.4969974874, 0.5030150754]$	0.0060175879
100	$[0.4999754000, 0.5000255013]$	0.0000510013
1000	$[0.4999997495, 0.5000002506]$	0.0000005011

Remark 2.15 It can be seen from the above example that there is little to be gained by going to $k > 1$ for small $w(X)$. This phenomenon is not an exception. It is connected to the fact that for every k, the quotient $w[F_k(X)]/w[F_1(X)]$ converges to 1 if X converges to c, as long as it is assumed that $F_1(X)$ exists, c is the centre of X, and $w[F_1(X)] \neq 0$ for all the intervals X. This property can easily be proved by writing down the quotient of the two widths, inserting the formulas for the widths, multiplying the numerator and the denominator with

$$|q(c)| - \sum_{|\lambda|=1}^{n} |D^\lambda q(c)| z^\lambda/\lambda!,$$

and cancelling as much as possible. An expression of the following form is therefore obtained,

$$\frac{w[F_k(X)]}{w[F_1(X)]} = \frac{q(c) \sum_{|\lambda|=r} |D^\lambda f(c)| z^\lambda/\lambda! + \sum_{|\lambda|>r} u_\lambda z^\lambda}{q(c) \sum_{|\lambda|=r} |D^\lambda f(c)| z^\lambda/\lambda! + \sum_{|\lambda|>r} v_\lambda z^\lambda},$$

where $u_\lambda, v_\lambda \in R$ and where r is an integer with $1 \leq r \leq n$, such that there exists a multi-index λ_0 with $|\lambda_0| = r$ and $D^{\lambda_0}f(c) \neq 0$. The existence of λ_0 results from the assumption $w[F(X)] \neq 0$. Remark 3.6 furthermore guarantees that $q(c) \neq 0$. Now $w(X) \to 0$ means $z \to o$, and the limiting process can therefore be carried out in the usual way.

2.6 KRAWCZYK'S CENTRED FORMS

The standard centred form is an explicit formula for an inclusion of the range $\tilde{f}(X)$ for a rational function $f = p/q$ over X. If the dimension of X is one, then the computational complexity, that is the number of arithmetic operations, of the standard centred form is $O(n^2)$, where n is the maximum of the degrees of the polynomials p and q. Krawczyk (1983) found a centred form that needs only $O(n)$ arithmetic operations for the just mentioned case. The gist of his centred form is the dependence on a *function procedure and the use of interval slopes*. The advantage of Krawczyk's form is a low computational complexity. This is contrasted by some disadvantages. The first disadvantage is that an explicit formula for Krawczyk's form cannot be given in general. Further, the form manipulates not only the symmetric interval $X - m(X)$, but also the unsymmetric interval X such that the general symmetry is lost. This implies that evaluations of Krawczyk's form are best done using a computer where interval arithmetic is implemented. The quadratic convergence of this form is proven in Section 3.4. The content of this section is due to Krawczyk (1983).

It is first necessary to introduce the concept of a function procedure. This concept was already used by Chuba–Miller (1972) and Miller (1972) for proving quadratic convergence properties of centred forms.

Definition 2.4 Let $B = \{b_1,\ldots,b_r\}$ be a finite set of real constants, $x = (x_1,\ldots,x_m)$ a variable over R^m and $U = \{u_1,\ldots,u_s\}$ a finite set of (dependent) variables over R. Then the finite sequence of instructions

$u_i := x_i$ for $i = 1,\ldots,m,$

$u_i := b_{i-m}$ for $i = m + 1,\ldots,m + r,$

$u_i := u_j *_i u_k$ for some $j, k < i$ and

$*_i \in \{+, -, \cdot, /\}$ for $i = m + r + 1,\ldots,s$

is called a *function procedure* S.

It is obvious that for any $i \in \{1,\ldots,s\}$, the first i instructions of S define a function $f_i = f_i(x)$ on a domain $D_i \subseteq R^m$ on which the functions $f_1,\ldots,f_i$ do not give rise to forbidden divisions through zero. We call f_i the *function corresponding to* u_i or *defined by the ith instruction* of S and $f = f_s$ the *function of the procedure* S. If the variable x is replaced by an interval variable $X = (X_1,...,X_m) \in I^m$ in the function procedure S, then we call the resulting sequence of instructions an *interval function procedure*. Again, the first i instructions of it determine an interval function $F_i = F_i(X)$ on a subset of I^m on which no forbidden division occurs. As is the case with expressions, functions and interval functions can be defined by arbitrarily many different procedures.

Example 2.9 If $m = r = 1$ and $B = \{1\}$ then the function procedure

$$u_1 := x(=x_1), \quad u_2 := 1, \quad u_3 := u_1 + u_2, \quad u_4 := u_1 u_3, \quad u_5 := u_1 + u_4$$

defines stepwise the functions

$$f_1(x) = x, f_2(x) = 1, f_3(x) = x + 1, f_4(x) = x(x + 1),$$
$$f_5(x) = x + x(x + 1) = x^2 + 2x.$$

The corresponding interval function procedure produces stepwise the interval functions

$$F_1(X) = X, F_2(X) = 1, F_3(X) = X + 1, F_4(X) = X(X + 1),$$
$$F_5(X) = X + X(X + 1).$$

If now f is a rational function in $x \in R^m$ and if $X \in I^m$ is an interval lying in the domain of f, then an interval $G \in I^m$ satisfying

$$f(x) - f(c) \in G \cdot (x - c) \quad \text{for all } x \in X \tag{2.36}$$

where $c = m(X)$ or any other point of X, is called an *interval slope of f in X*. Krawczyk's method now consists of the construction of an interval function procedure for an interval slope $G \in I^m$ of f in X such that the relation (2.36) can be subordinated to Moore's concept of a centred form (see Section 2.1).

Let $e_i \in I^m$ denote the ith unit-vector and $o \in R^m$ the null-vector, let $X \in I^m$ and $c = m(X)$ or any other point of X. Operations of the form AB or A/B where $A \in I^m$ and $B \in I$ are to be performed component-wise in this section. Contrary to our conventions, G_i does not denote the ith component of the interval vector G in this section, but autonomous interval vectors of I^m.

Definition 2.5 Let a function procedure S be given as in Definition 2.4. Then the *interval slope* $G = G_s \in I^m$ of S over X is recursively defined by

$G_i = e_i$ for $i = 1,...,m$,

$G_i = o$ for $i = m + 1,..., m + r$,

$\left. \begin{array}{l} G_i = G_j \pm G_k \text{ if } u_i := u_j \pm u_k, \\ G_i = G_j F_k(X) + G_k f_j(c) \text{ if } u_i := u_j u_k, \\ G_i = G_j/F_k(X) - G_k(X)f_j(c)/F_k(X) \text{ if } u_i := u_j/u_k \end{array} \right\}$ for $i = m + r + 1,...,s$, if no forbidden division occurs.

We show next that the interval slope G of S is also an interval slope of the function f defined by S. Therefore G can be used for including the range $\bar{f}(X)$ via (2.36). Slopes for bounding the range of polynomials were already used by Alefeld (1981).

Lemma 2.,8 *Let f be the function of the function procedure S. If the interval slope G of S is defined over X, then G is an interval slope of f over X.*

Proof We show by induction that

$$G_i \text{ is an interval slope of } f_i \text{ over } X \qquad (2.37)$$

for $i = 1,\ldots,s$. Then the assertion is proved. First, the validity of (2.37) is evident for $i = 1,\ldots,m + r$. Let us now assume that $m + r + 1 \leq i \leq s$ and that (2.37) holds for $j < i$ and $k < i$. Then, in case of addition and subtraction,

$$f_i(x) - f_i(c) = f_j(x) - f_j(c) \pm (f_k(x) - f_k(c))$$
$$\in G_j \cdot (x - c) \pm G_k \cdot (x - c)$$
$$= G_i \cdot (x - c).$$

The last rearrangement is done by Lemma 1.3. If $u_i := u_j u_k$, we get

$$f_i(x) - f_i(c) = (f_j(x) - f_j(c)) f_k(x) + (f_k(x) - f_k(c)) f_j(c)$$
$$\in (G_j F_k(X) + G_k f_j(c)) \cdot (x - c)$$
$$= G_i \cdot (x - c).$$

If $u_i := u_j/u_k$, we get

$$f_i(x) - f_i(c) = (f_j(x) - f_j(c))/f_k(x) - f_j(c)(f_k(x) - f_k(c))/$$
$$(f_k(x) f_k(c))$$
$$\in (G_j/F_k(X)) \cdot (x - c) - (G_k f_i(c)/F_k(X)) \cdot (x - c)$$
$$= G_i \cdot (x - c). \qquad \square$$

We are now ready for the final definition.

Definition 2.6 *Let f be the function of the function procedure S and G the interval slope of S over $X \in I^m$. Then the interval*

$$F(X) = f(c) + G \cdot (X - c)$$

is called Krawczyk's centred form.

It follows from (2.36) and Lemma 2.8. that $F(X) \supseteq \bar{f}(X)$. We will see that $G = G(X)$ satisfies a Lipschitz condition and hence, that $F(X)$ is quadratically convergent if X is seen as an interval variable over a compact domain in Section 3.4.

The computational complexity of Krawczyk's centred form is very low compared to the standard centred form. Since $k = s - m - r$ arithmetic operations are needed for describing the function procedure S, only $O(k)$ interval operations are needed for the computation of the interval slope G and of the form $F(X)$. If for example, $f = p/q$ is a function in one variable $x \in R$ and n is the maximum degree of the polynomials p and q, then one can find a function procedure S for f which has $5n$ arithmetic operations. Hence $O(n)$ interval operations are necessary for the calculation of $F(X)$.

Krawczyk (1983) compares several kinds of function procedures with respect to the quality of the resulting centred form.

Example 2.10 Let $f(x) = 1/x^2$ be defined on the interval $X = [2, 4]$. Then

$$w[\tilde{f}(X)] = 0.18750.$$

Two different function procedures for f shall be considered:

The function procedure S shall be given by $u_0 := x$, $u_1 := 1$, $u_2 := u_0 u_0$, $u_3 := u_1/u_2$ and leads to the interval slope $G = -[5/144, 7/36]$ and to a centred form $F(X)$ with

$$w[F(X)] = 7/18 = 0.38889.$$

Let another function procedure S be given by $u_0 := x$, $u_1 := 1$, $u_2 := u_1/u_0$, $u_3 := u_2/u_0$. It leads to an interval slope $G = -[7/144, 5/36]$ and to a centred form $F(X)$ with

$$w[F(X)] = 5/18 = 0.27778.$$

Comparing these results with the standard centred forms $\tilde{F}_k(X)$ calculated with extended arithmetic we get (see Ratschek, 1980a):

k	$w[\tilde{F}_k(X)]$
1	0.48148
2	0.30864
3	0.29630
4	0.24966
5	0.24371

CHAPTER III

General definition of centred forms

In Chapter II we treated the standard centred forms for rational functions. They are easy to manipulate and easy to program. Furthermore, they have been applied to practical problems of various kinds. Presently we will introduce and discuss an axiomatic definition of centred forms valid also for real functions. This definition has its origin in the work of Krawczyk–Nickel (1982). A number of considerations relating to centred forms are also pursued.

3.1 THE REASONS FOR A GENERAL DEFINITION

The definitions of standard centred forms given in Chapter II are adequate for many purposes. In some instances, however, these definitions do not suffice. A general definition of centred forms by axioms as given in the sequel will have the following advantages:

(a) The centred form method can be extended to real, complex, vector valued functions, etc. via the axioms.
(b) The mean value form turns out to be a special centred form (Krawczyk–Nickel (1982) and Section 3.5). The theories of both forms which have up to now been developed separately may therefore be represented in an unified manner.
(c) It is possible to generate various concrete centred forms depending on the actual purpose and the information available.
(d) The general definition is not given by explicit and complicated formulas but implicitly by few characteristic arithmetic and analytical properties which are easy to deal with. It is therefore an appropriate and convenient basis for theoretical investigations.
(e) Krawczyk–Nickel (1982) give an important characterization of a class of centred forms which are inclusion isotone a property which facilitates the subdivision method given in Chapter IV.

(f) The proof of the quadratic convergence is elegant and can be executed in very general normed or metric interval spaces (Krawczyk–Nickel, 1982).

Since the general definition given in Section 3.2 is a framework for theoretical investigations rather than an explicitly applicable formula we do not insist that $c = m(X)$. For actual forms satisfying the axioms the choice $c = m(X)$ is again recommended by us. The reasons are the same as discussed in Section 2.3. Some examples of centred forms for non-rational functions will also be given. We will devote some space to these examples since, as far as we know, there are no such examples available in the literature.

The general definition is discussed in interval and normed spaces in Krawczyk–Nickel (1982) in order to obtain convergence theorems that are as general as possible. It is shown, however, in the Appendix that all continuous norms on I^m are equivalent. Since there are close connections between the width of intervals and the maximum norm (see Section 1.3), a definition using norms and special metrics such as the definition of Krawczyk–Nickel (1982) may therefore be replaced by an equivalent definition using only the width. This is the approach followed here.

3.2 GENERAL DEFINITION OF CENTRED FORMS

In the definition of the standard centred forms the underlying domain X was a constant interval (or parallelepiped). The concept of quadratic convergence will only make sense if we let the domain vary (shrink). The domain is therefore now considered to be an interval variable $Y \subseteq X$. It also follows from this that the developing point c is in general a variable. The functional relationship between the domain Y and the developing point c is represented by a function α. If the reader is less interested in quadratic convergence than in obtaining an inclusion for the range it is possible to suppress the use of the variable Y as well as the function α.

We consider *continuous* functions $f: D \to R$ with $D \subseteq R^m$. The notation $X = X_1 \times \ldots \times X_m = (X_1, \ldots, X_m) \in I(D)$ means a right parallelepiped with $X_i \in I$ and $X \subseteq D$. This is also called an *interval*. If $X_i = [x_i, y_i]$ then X is also represented by $[x, y] = [(x_1, \ldots, x_m), (y_1, \ldots, y_m)]$. Furthermore, $c = (c_1, \ldots, c_m)$ always denotes a point of D, and $H = X - c$ is known as the 'centring' of X. The dot '·' denotes the inner product $R^m \times R^m \to R$ in the usual manner. It can also be extended to the case

$$R^m \times I^m \to I \quad \text{or} \quad I^m \times I^m \to I.$$

For instance, if $H = (H_1, \ldots, H_m) \in I^m$ and $G = (G_1, \ldots, G_m) \in I^m$ then we have

$$G \cdot H = G_1 H_1 + \ldots + G_m H_m.$$

Sec. 3.2]   **General Definition of Centred Forms**   65

Definition 3.1 Let f, D, and X be defined as above, $D_0 \subseteq D$, and let $\alpha: I(X) \to D_0$ be a function. The function $s: X \times D_0 \to R$ is defined by $s(x, c) = f(x) - f(c)$ for all $x \in X$, $c \in D_0$. If there exists a positive integer r and interval functions $S: I(X) \to I$ and $G^\rho: I(X) \to I^m$, $\rho = 1,\ldots,r$, such that

$$s(x, \alpha(Y)) \in S(Y) \subseteq \sum_{\rho=1}^{r} (Y - \alpha(Y)) \cdot G^\rho(Y) \tag{3.1}$$

for all $Y \in I(X)$, $x \in Y$, then the interval function $F: I(X) \to I$, defined by

$$F(Y) = f[\alpha(Y)] + S(Y) \tag{3.2}$$

is called *a centred form function* of f on X with *a developing point function* α. If $Y \in I(X)$, $c = \alpha(Y)$, then the interval $F(Y) = f(c) + S(Y)$ is called *a centred form* of f on Y and c its *developing point*.

The difference between a centred form and a centred form function is normally not considered. Definition 3.1 is rather complicated. It deviates from the usual definitions and an extensive discussion of this definition will therefore be given in the next section. In the present section we only give some simple facts as well as an example which will supplement Definition 3.1.

The desired inclusion property

$$\tilde{f}(X) \subseteq F(X) \tag{3.3}$$

is valid. This follows directly from (3.1). It can be written more generally in the form

$$\tilde{f}(Y) \subseteq F(Y) \quad \text{for all } Y \in I(X). \tag{3.4}$$

A function $F: I(X) \to I$ which satisfies (3.4) will be called an *inclusion function* for $\tilde{f}$ over X.

Remark 3.1 In our definition we only consider functions f with values in R. That is not a restriction since in the case of vector valued functions, $f = (f_1,\ldots,f_k)$ with $f_k: D \to R$, and $D \subseteq R^m$, the definitions and theorems can be applied componentwise, that is, to each component $f_1,\ldots,f_k$, in order to get the assertions for f.

A simpler definition of the general centred form is given for readers who are not interested in quadratic convergence.

Definition 3.2 Let f, D, and X be defined as above. Let $c \in D$ and $H = X - c$. The function $s: X \to R$ is defined by $s(x) = f(x) - f(c)$ for all $x \in X$. If there exists a positive integer r and intervals $S \in I$ and $G^1,\ldots,G^r \in I^m$ such that

$$s(x) \in S \subseteq \sum_{\rho=1}^{r} H \cdot G^\rho,$$

then the interval $F(X) = f(c) + S$ is called a *centred form* with c as developing point.

It can be verified at once that Definition 3.2 is a special case of Definition 3.1 by considering s as independent of the former variable c, and the interval functions S and G^ρ as independent of the interval variable $Y \subseteq X$.

3.3 THE QUADRATIC CONVERGENCE

The quadratic convergence of a centred form function tells us that the centred form function converges quadratically to the range function if the width of the domain converges to 0. Since the centred form methods and the quadratic convergence strongly influence each other we prove the main results on quadratic convergence in this section and continue the general discussion of quadratic convergence in Chapter IV.

An inclusion function $F: I(X) \to I$ for $\tilde{f}$ is called *linearly convergent* (to the range function $\tilde{f}$) if there exists a real number K such that

$$w[F(Y)] - w[\tilde{f}(Y)] \leq Kw(Y) \quad \text{for all } Y \in I(X).$$

Furthermore, F is called *quadratically convergent* (to $\tilde{f}$) if there exists a real number K such that

$$w[F(Y)] - w[\tilde{f}(Y)] \leq Kw(Y)^2 \quad \text{for all } Y \in I(X).$$

The quadratic convergence property was first conjectured by Moore (1966) and, for rational functions in one and several variables, first proven by Hansen (1969b). A very interesting proof for rational functions in several variables was developed by Chuba–Miller (1972), Miller (1972). This proof uses simple recursive steps. A further proof of quadratic convergence is given by Alefeld–Herzberger (1974, 1983). The most general proof which we follow in the present monograph is due to Krawczyk–Nickel (1982). This proof formulates the quadratic convergence theorems in very general normed spaces using several metrics and seems to be the first proof which is also valid for non-rational functions. In this monograph we simplify the proofs by using the width of intervals in describing the convergence theorems. This is not a restriction from the mathematical point of view as discussed in Section 3.1. We do, however, use the Hausdorff metrics and norms in the *proof* of the convergence theorem in order to keep the proof as simple as possible.

The key to the convergence proof is provided by the following two lemmas:

Lemma 3.1 (*Miranda*, 1941). *Let* $H = (H_1,\ldots,H_m) \in I^m$ *and* $H_i = [u_i, v_i]$. *Let the function* $k = (k_1,\ldots,k_m): H \to R^m$ *be continuous and satisfying the m* sign conditions $(i = 1,\ldots,m)$,

$$k_i(h_1,\ldots,h_{i-1}, u_i, h_{i+1},\ldots,h_m) \leq 0,$$
$$k_i(h_1,\ldots,h_{i-1}, v_i, h_{i+1},\ldots,h_m) \geq 0.$$

Then a zero of k exists in H. □

The proof depends on the Brouwer's fixed point theorem and will not be repeated here.

Lemma 3.2 (Krawczyk–Nickel, 1982). *Let $H = (H_1,\ldots,H_m) \in I^m$, $o \in H$, and let the function $g = (g_1,\ldots,g_m): H \to R^m$ be continuous. If the functions $\varphi: H \to R$ and $\Phi = [\varphi_1, \varphi_2]: H \to I$ are defined by*

$$\varphi(h) = h \cdot g(h) = \sum_{i=1}^{m} h_i g_i(h),$$

$$\Phi(h) = H \cdot g(h) = \sum_{i=1}^{m} H_i g_i(h),$$

then there exist two vectors $h_*, h^* \in H$ that satisfy

$$\varphi(h_*) = \varphi_1(h_*) \quad \text{and} \quad \varphi(h^*) = \varphi_2(h^*).$$

Proof It is sufficient to prove the existence of h^*, since the existence of h_* follows by symmetry. Let the function $k: H \to R^m$ be defined componentwise by

$$k_i(h) = \begin{cases} (h_i - v_i)g_i(h) & \text{if } g_i(h) \geq 0 \\ (u_i - h_i)g_i(h) & \text{if } g_i(h) \leq 0, \end{cases}$$

where $H_i = [u_i, v_i]$. The continuity of k follows from the continuity of g. From the definition of k_i we have

$$k_i(h) \leq 0 \text{ if } h_i = u_i,$$
$$k_i(h) \geq 0 \text{ if } h_i = v_i.$$

This means that the sign conditions of Lemma 3.1 hold which implies that there exists a zero $\hat{h}$ of k in H. Using the relation

$$\max(H \cdot g(\hat{h})) = \sum \max\{u_i g_i(\hat{h}), v_i g_i(\hat{h})\}$$

as well as the relation

$$\min\{(\hat{h}_i - u_i) g_i(\hat{h}), (\hat{h}_i - v_i) g_i(\hat{h})\} = k_i(\hat{h})\operatorname{sgn} g_i(\hat{h})$$

which follows from the assumptions $u_i \leq 0 \leq v_i$ as well as the definition of

$k_i(\hat{h})$ we get

$$\varphi(\hat{h}) - \varphi_2(\hat{h}) = \hat{h} \cdot g(\hat{h}) - \max(H \cdot g(\hat{h}))$$

$$= \sum_{i=1}^{m} [\hat{h}_i g_i(\hat{h}) - \max\{u_i g_i(\hat{h}), v_i g_i(\hat{h})\}]$$

$$= \sum_{i=1}^{m} \min\{(\hat{h}_i - u_i) g_i(\hat{h}), (\hat{h}_i - v_i) g_i(\hat{h})\}$$

$$= \sum_{i=1}^{m} k_i(\hat{h}) \operatorname{sgn} g_i(\hat{h}) = 0. \qquad \square$$

If $Y \in I^m$ and if $c \in R^m$ then we recall the notation $Y \vee c$ for the smallest interval in I^m containing both Y and c (see also Section 1.3).

Lemma 3.3 (*Extension Lemma*). *Let G, $Y \in I^m$, $c \in R^m$, and let $g: Y \to G$ be continuous. Then there exists a continuous extension $\tilde{g}: (Y \vee c) \to G$ of g.*

Proof If $c \in Y$, then we do not have to prove anything. We therefore assume $c \notin Y$. If $\| \ \|_E$ denotes the Euclidian norm on R^m, then there exists a nearest point $\hat{u} \in Y$ to each point $u \in Y \vee c$. This means that

$$\|u - \hat{u}\|_E \leq \|u - x\|_E \qquad \text{for all } x \in Y.$$

The point $\hat{u}$ is uniquely determined and, if the components of Y are $Y_i = [a_i, b_i]$, then we have

$\hat{u}_i = u_i$ if $u_i \in Y_i$

$\hat{u}_i = b_i$ if $u_i > b_i$

$\hat{u}_i = a_i$ if $u_i < a_i$.

The mapping $u \to \hat{u}$ is continuous since the component functions $u \to \hat{u}_i$ are continuous. The extension $\tilde{g}$ of g is then defined by

$$\tilde{g}(u) = g(\hat{u}) \qquad \text{for } u \in Y \vee c.$$

The function g is continuous since it is the composition of continuous functions. Since $\hat{u} \in Y$ we have $\tilde{g}(u) = g(\hat{u}) \in G$. $\qquad \square$

We are now ready to state the main theorems on linear and quadratic convergence of the centred form functions. The following theorem has its origin in Moore (1966).

Theorem 3.1 *Let a centred form function F of a function f be defined as in Definition 3.1. If the functions $G^\rho: I(X) \to I^m$, $\rho = 1, \ldots, r$, are bounded and if the developing point function $\alpha: I(X) \to D_0$ satisfies $\alpha(Y) \in Y$ for all $Y \in I(X)$, then F is linearly convergent to the range function $\bar{f}$.*

Proof Suppose that K is a bound for the functions G^ρ. The formulas (1.3) and (1.6) are used for the following estimation valid for $Y \in I(X)$:

$$w[F(Y)] - w[\tilde{f}(Y)] \leq w[F(Y)] = w[S(Y)]$$

$$\leq \sum_{\rho=1}^{r} w[(Y - \alpha(Y)) \cdot G^\rho(Y)]$$

$$= \sum_{\rho=1}^{r} \sum_{\mu=1}^{m} w[(Y_\mu - \alpha_\mu(Y)) G_\mu^\rho(Y)]$$

$$\leq \sum_{\rho=1}^{r} \sum_{\mu=1}^{m} [w(Y_\mu - \alpha_\mu(Y)) |G_\mu^\rho(Y)| + |Y_\mu - \alpha_\mu(Y)| w(G_\mu^\rho(Y))]$$

$$\leq \sum_{\rho=1}^{r} \sum_{\mu=1}^{m} [w(Y_\mu - \alpha_\mu(Y))K + w(Y_\mu - \alpha_\mu(Y)) 2K]$$

$$\leq rm[w(Y)K + w(Y)2K] = 3rmKw(Y). \qquad \square$$

The following lemma is due to Krawczyk–Nickel (1982) and contains the crucial estimation of the width-difference $w[F(Y)] - w[\tilde{f}(Y)]$. It should be pointed out that in this lemma only a fixed interval $Y \in I(X)$ is considered. Relationships depending on all $Y \in I(X)$ are first encountered in Theorem 3.2.

Lemma 3.4 *Let $D \subseteq R^m$, $c \in D$, $Y \in I(D)$, $G \in I^m$, and let the function $f: D \to R$ be continuous in Y and representable on the form*

$$f(x) = f(c) + (x - c) \cdot g_c(x) \qquad \text{for all } x \in Y \tag{3.5}$$

where $g_c: Y \to G$ is a suitable continuous function. If a centred form of f on Y is defined by

$$F(Y) = f(c) + (Y - c) \cdot G$$

then

$$w[F(Y)] - w[\tilde{f}(Y)] \leq 2mw(G)\|Y - c\|. \tag{3.6}$$

Proof First we extend g_c to a continuous function $g_c: (Y \vee c) \to G$ by Lemma 3.3. Then we introduce the following abbreviations and notations:

$h = x - c \qquad$ for $x \in Y \vee c$,

$Y = [y_1, y_2]$,

$H = (Y - c) \vee o$,

$g(h) = g_c(h + c) = g_c(x) \qquad$ for $h \in H$ respectively $x \in Y \vee c$,

$\varphi(h) = h \cdot g(h) \qquad$ for $h \in H$,

$\Phi(h) = H \cdot g(h) = [\varphi_1(h), \varphi_2(h)] \qquad$ for $h \in H$,

$\tilde{\varphi}(H) = \{\varphi(h): h \in H\}$.

Now, we have

$$\varphi(h) \in \bar{\varphi}(H) \subseteq H \cdot G \quad \text{for all } h \in H,$$
$$F(Y) \subseteq f(c) + H \cdot G,$$
$$\tilde{f}(Y) = f(c) + \bar{\varphi}(H).$$

We use the Hausdorff metric for brevity. Applying its chain inclusion isotonicity to the chain

$$\tilde{f}(Y) \subseteq F(Y) \subseteq f(c) + H \cdot G$$

and then using the translation invariance we get

$$|F(Y), \tilde{f}(Y)| \leq |H \cdot G, \bar{\varphi}(H)|. \tag{3.7}$$

Since g_c and therefore g is continuous using Lemma 3.2, and the fact that $0 \in \Phi(h)$ for $h \in H$ we get

$$\varphi(h_*) = \varphi_1(h_*) \leq 0,$$
$$\varphi(h^*) = \varphi_2(h^*) \geq 0.$$

Accordingly we have

$$\Phi(h_*) \cup \Phi(h^*) = [\varphi(h_*), \varphi_2(h^*)] \cup [\varphi_1(h^*), \varphi(h^*)]$$
$$= [\max\{\varphi(h_*), \varphi_1(h^*)\}, \min\{\varphi_2(h_*), \varphi(h^*)\}] \subseteq$$
$$\subseteq [\varphi(h_*), \varphi(h^*)] \subseteq \bar{\varphi}(H).$$

We can therefore apply Lemma 1.5 to the following chain,

$$\Phi(h_*) \cap \Phi(h^*) \subseteq \bar{\varphi}(H) \subseteq H \cdot G,$$

since the intersection is non-empty, see (3.7), and we get

$$|H \cdot G, \bar{\varphi}(H)| \leq \max\{|H \cdot G, \Phi(h_*)|, |H \cdot G, \Phi(h^*)|\}. \tag{3.8}$$

In order to make the distances occurring in (3.8) more transparent we use (1.13) and get for any $h \in H$,

$$|H \cdot G, \Phi(h)| = |H \cdot G, H \cdot g(h)| \leq m\|H\| |G, g(h)|.$$

Inserting this inequality in (3.8) eliminating the terms in which the Hausdorff-metrics occur by (1.10) and using the identity $\|H\| = \|Y - c\|$, we get

$$w[F(Y)] - w[\tilde{f}(Y)] \leq 2|F(Y), \tilde{f}(Y)|$$
$$\leq 2m\|Y - c\| [w(G) - w(g(H))] = 2m\|Y - c\| w(G),$$

using the equation (3.7) as well. □

We will now prepare the assumptions for the Theorem 3.2.

Let $D \subseteq R^m$, $X \in I(D)$, and let $\alpha: I(X) \to D_0$ be bounded, where $D_0 \subseteq D$. Let the continuous function $f: D \to R$ be represented on X in the form

$$f(x) = f(c) + (x - c) \cdot g(x, c) \qquad \text{for all } c \in D_0, x \in X \tag{3.9}$$

where $g: X \times D_0 \to R^m$ is a suitable mapping continuous in the first variable (over X). Let the interval functions $G: I(X) \to I^m$ and $S: I(X) \to I$ be defined such that for all $Y \in I(X)$ and $x \in Y$ the following two conditions

$$g(x, \alpha(Y)) \in G(Y), \tag{3.10}$$

$$(x - \alpha(Y)) \cdot g(x, \alpha(Y)) \in S(Y) \subseteq (Y - \alpha(Y)) \cdot G(Y) \tag{3.11}$$

are valid. Clearly the function $F: I(X) \to I$ defined by

$$F(Y) = f(\alpha(Y)) + S(Y) \qquad \text{for all } Y \in I(X)$$

is a centred form function for f in X. The following theorem contains the main results of linear and quadratic convergence of F to $\tilde{f}$ and is applicable to many convergence problems with arise in connection with centred forms. The notation and assumptions which have just been introduced are assumed to hold.

Theorem 3.2 (*Krawczyk–Nickel, 1982*). *If the function G is Lipschitz, then the centred form function F converges linearly to the range function $\tilde{f}$. If further $\alpha(Y) \in Y$ for all $Y \in I(X)$ then the convergence is quadratic.*

Proof The assertions follow directly from Lemma 3.4. We only have to keep in mind that the intervals G and Y are fixed in Lemma 3.4. In the present theorem, however, Y acts as a variable, and a certain G corresponds to each $Y \in I(X)$. It is natural to handle this relationship as a function $Y \to G(Y)$.

Since G is a Lipschitz function, there exists $\lambda \in R$ such that

$$w[G(Y)] \leq \lambda w(Y) \qquad \text{for all } Y \in I(X),$$

and, by Lemma 3.4, we have

$$w[F(Y)] - w[\tilde{f}(Y)] \leq 2m\lambda w(Y) \|Y - \alpha(Y)\|. \tag{3.12}$$

Since α is bounded and $Y \subseteq X$ there exists $K \in R$ such that

$$\|Y - \alpha(Y)\| \leq K \qquad \text{for all } Y \in I(X).$$

This means that the linear convergence is obtained, i.e.,

$$w[F(Y)] - w[\tilde{f}(Y)] \leq 2m\lambda K w(Y).$$

If $\alpha(Y) \in Y$, then

$$\|Y - \alpha(Y)\| \leq w(Y), \tag{3.13}$$

and the quadratic convergence follows from (3.12), that is

$$w[F(Y)] - w[\tilde{f}(Y)] \leq 2m\lambda w(Y)^2. \qquad \square \qquad (3.14)$$

Remark 3.2 (Krawczyk–Nickel, 1982). The estimation (3.14) of previous proof can be improved by 50%, that is,

$$w[F(Y)] - w[\tilde{f}(Y)] \leq m\lambda w(Y)^2$$

if α is the midpoint function, since in this case (3.13) can be replaced by $\|Y - \alpha(Y)\| = w(Y)/2$.

Remark 3.3 It is also possible to express the Lipschitz condition for G in terms of a homogeneous, translation invariant, chain inclusion isotonic metric (for a definition see Section 1.3), see, for instance, Krawczyk–Nickel (1982), Raith–Rokne (1982).

3.4 THE STANDARD AND KRAWCZYK'S CENTRED FORMS

The standard centred form for a rational function $f = p/q$ of any order k, in one or several variables, is a centred form where the function G satisfies the Lipschitz condition. The developing point function is $\alpha(Y) = m(Y)$ where $D_0 = X$. We first look at the case of one variable. We obtain

$$s(x, c) = \sum_{\lambda=1}^{k-1} f^{(\lambda)}(c) h^\lambda/\lambda! + \left(\sum_{\lambda=1}^{k+n-1} t_{k\lambda} h^\lambda/\lambda! \right) \bigg/ \left[\sum_{\lambda=0}^{n} q^{(\lambda)}(c) h^\lambda/\lambda! \right]$$

for all $x, c \in X$, where $h = x - c$. It is understood that the coefficients $t_{k\lambda}$ that occur in the above expression depend on the variable c as expressed by (2.19). In order to construct the interval function S we consider, for given $Y \in I(X)$, the function value $S(Y)$. Comparing the function s to Definition 2.1, we recognize that $S(Y)$ is nothing but the natural interval extension of $s(x, \alpha(Y))$ to Y where x is replaced by Y and where h is replaced by $Z = Y - \alpha(Y)$. We therefore obtain

$$S(Y) = \sum_{\lambda=1}^{k-1} f^{(\lambda)}(c) Z^\lambda/\lambda! + \left(\sum_{\lambda=k}^{k+n-1} t_{k\lambda} Z^\lambda/\lambda! \right) \bigg/ Q(Y) \qquad (3.15)$$

where $Q(Y)$ is the standard centred form of q on x of first order if $f = p/q$ and where $c = \alpha(Y)$. It is now easy to check that the following formula holds for intervals $A \in I$ and symmetric intervals $U, V \in I$,

$$U(A + V) = UA + UV. \qquad (3.16)$$

Applying (3.16) to (3.15), we get

$$S(Y) = ZG(Y)$$

Sec. 3.4] The Standard and Krawczyk's Centred Forms

where

$$G(Y) = \sum_{\lambda=1}^{k-1} f^{(\lambda)}(c) Z^{\lambda-1}/\lambda! + \left(\sum_{\lambda=k}^{k+n-1} t_{k\lambda} Z^{\lambda-1}/\lambda! \right) / Q(Y).$$

The function S (or G) exists if $F_k(X)$ exists which is assumed.

We remark that the proof that F_k is a centred form is easy since $Z = Y - \alpha(Y)$ is symmetric which yields $r = 1$ and $S = ZG$, where $G = G^1$, see Definition 3.1.

In case of several variables the procedure is similar. We have nevertheless separated the two cases since in the present consideration it is more troublesome to find the function $G = G^1$.

By (2.29) and Lemma 2.5 we get

$$s(x, c) = f(x) - f(c) = \sum_{|\lambda|=1}^{k-1} D^\lambda f(c) h^\lambda/\lambda!$$

$$+ \left(\sum_{|\lambda|=k}^{k+n-1} t_{k\lambda} h^\lambda/\lambda! \right) \Big/ \left(\sum_{|\lambda|=0}^{n} D^\lambda q(c) h^\lambda/\lambda! \right) \quad \text{for } x, c \in X$$

where $h = x - c$ and where the coefficients $t_{k\lambda}$ depend on the variable c as in the previous case.

First, we remember the notation $h = (h_1,\ldots,h_m)$, $Z = (Z_1,\ldots,Z_m)$, etc. Then we denote the ith unit-vector of the space R^m by $e_i = (0,\ldots,1,\ldots,0)$. Furthermore, we introduce the notation $\Phi_i(\lambda)$ for multi-indices λ as abbreviation for the condition

$$\lambda_1 = \ldots = \lambda_{i-1} = 0, \lambda_i > 0 \quad (i = 1,\ldots,m).$$

We notice that $\bigcup_{i=1}^{m}\{\lambda: \Phi_i(\lambda)\}$ is a disjunct partition of the set of all multi-indices λ.

The function s can now be written as

$$s(x, c) = \sum_{i=1}^{m} s_i(x, c) \quad \text{for } x \in X, c \in X$$

where

$$s_i(x, c) = h_i g_i(x, c)$$

and

$$g_i(x, c) = \sum_{\substack{|\lambda|=1 \\ \Phi_i(\lambda)}}^{k-1} D^\lambda f(c) h^{\lambda-e_i}/\lambda! + \sum_{\substack{|\lambda|=k \\ \Phi_i(\lambda)}}^{k+n-1} (t_{k\lambda} h^{\lambda-e_i}/\lambda!) \Big/ \left(\sum_{|\mu|=k}^{k+n-1} D^\mu q(c) h^\mu/\mu! \right).$$

Defining $g = (g_1,\ldots,g_m): X \times X \to R^m$ we may write $s(x, c) = h \cdot g(x, c)$. In order to be precise we now consider the previous equation as being dependent on the intervals $Y \in I(X)$. Furthermore, we only consider the special case $c = \alpha(Y)$. This results in

$$s(x, \alpha(Y)) = (x - \alpha(Y)) \cdot g(x, \alpha(Y))$$

for all $Y \in I(X)$, $x \in Y$. If we now compare s with the definition of $F_k(X)$ in Definition 2.2 we realize that the interval $S(Y)$ is the natural interval extension of $s(x), \alpha(Y))$ to Y where x is replaced by Y and h by $Z = Y - \alpha(Y)$. We therefore obtain

$$S(Y) = \sum_{|\lambda|=1}^{k-1} D^\lambda f(c) Z^\lambda/\lambda! + \left(\sum_{|\lambda|=k}^{k+n-1} t_{k\lambda} Z^\lambda/\lambda!\right)/Q(Y).$$

The interval extension of $g(x, \alpha(Y))$ to Y, that is, replacing x by Y, gives the interval $G(Y) = (G_1, \ldots, G_m)(Y) \in I^m$ where

$$G_i(Y) = \sum_{\substack{|\lambda|=1 \\ \Phi_i(\lambda)}}^{k-1} D^\lambda f(c) Z^{\lambda-e_i}/\lambda! + \left(\sum_{\substack{|\lambda|=k \\ \Phi_i(\lambda)}}^{k+n-1} t_{k\lambda} Z^{\lambda-e_i}/\lambda!\right)/Q(Y).$$

It remains to show that

$$S(Y) = Z \cdot G(Y) = \sum_{i=1}^{m} Z_i G_i(Y). \tag{3.17}$$

The proof is feasible since Z is symmetric and since the distributive law is applicable to perform the necessary rearrangements. In fact, we use (3.16) and the formula

$$A(U + V) = AU + AV \tag{3.18}$$

valid for arbitrary intervals $A \in I$ and symmetric intervals $U, V \in I$. The proof of (3.18) is simple and is left to the reader. In order to avoid involved formulas we replace the real coefficients in the previous formula for $S(Y)$ by a_λ and we set $A_\lambda = a_\lambda/Q(Y)$ whenever appropriate. With this we obtain:

$$\begin{aligned}
S(Y) &= \sum_{|\lambda|=1}^{k-1} a_\lambda Z^\lambda + \left(\sum_{|\lambda|=k}^{k+n-1} a_\lambda Z^\lambda\right)/Q(Y) \\
&= \sum_{|\lambda|=1}^{k-1} a_\lambda Z^\lambda + \sum_{|\lambda|=k}^{k+n-1} A_\lambda Z^\lambda \\
&= \sum_{i=1}^{m} \left[\sum_{\substack{|\lambda|=1 \\ \Phi_i(\lambda)}}^{k-1} a_\lambda Z^\lambda + \sum_{\substack{|\lambda|=1 \\ \Phi_i(\lambda)}}^{k+n-1} A_\lambda Z^\lambda\right] \\
&= \sum_{i=1}^{m} \left[Z_i \sum_{\substack{|\lambda|=1 \\ \Phi_i(\lambda)}}^{k-1} a_\lambda Z^{\lambda-e_i} + Z_i \sum_{\substack{|\lambda|=k \\ \Phi_i(\lambda)}}^{k+n-1} A_\lambda Z^{\lambda-e_i}\right] \\
&= \sum_{i=1}^{m} Z_i G_i(Y) = Z \cdot G(Y)
\end{aligned}$$

The second equality in the proof is valid by (3.18), the third one is only based on a rearrangement, the fourth and fifth one uses (3.16).

The symmetry of Z is again the reason for the simple proof. This means that $S(Y) = Z \cdot G(Y)$ is valid, and that it is sufficient to consider only one G^ρ (that is, $r = 1$).

Finally, it is easy to show that G satisfies a Lipschitz condition. We only have to show that each G_i satisfies a Lipschitz condition. The functions G_i are described in Theorem 1.2. First, we provide the continuous Lipschitz function $\varphi_1(Y) = m(Y)$ for the construction of G_i, see Theorem 1.2. Then the function value $G_i(Y)$ can be calculated from the argument Y, some constants (coefficients of f, etc.), and $\varphi_1(Y)$, using the four interval operations only. (Note that it is not possible to represent the function $Y \to m(Y)$ by an expression which uses only Y, constants, and the four operations. The reason for this is that in order to calculate $m(Y)$ we need not only the interval Y, but also the endpoints of Y explicitly.) The existence of the functions G_i (which is necessary for applying the theorem), that is the condition $0 \notin Q(Y)$, follows from the assumed existence of $F_k(Y)$ and the inclusion isotonicity of the standard centred form as in the case of one variable.

We now turn our attention to Krawczyk's centred form. If $X \in I^m$ is the basic interval then the functional version of this form is

$$F(Y) = f(c) + G(Y) \cdot (Y - c) \quad \text{for } Y \in I(X)$$

where $c = \alpha(Y)$ is the midpoint function or any other developing point function. It follows immediately from Definition 2.6 and Lemma 2.8 that this form is a special case of the general case presented in Definition 3.1. If $\alpha(Y) \in Y$ then Krawczyk's form is quadratically convergent by Theorem 3.2, since the interval slope $G(Y)$ is Lipschitz, which is a direct consequence of applying Theorem 1.2 to the recursive steps of Definition 2.5.

3.5 MEAN-VALUE AND TAYLOR-FORMS

In Moore (1966) it was suggested that the mean-value theorem may be used to obtain inclusions for the range of real functions over a parallelepiped. For the case of one real function of one real variable the suggestion was to first expand the function f around a point $c \in X$ obtaining

$$f(x) = f(c) + f'(c + \theta(x - c))(x - c)$$

for some $\theta \in [0, 1]$. Moore now assumed that f' had an inclusion F', that is $f'(x) \in F'(X)$ for $X \in I$ and all $x \in X$. We note that F' is not the derivative of F. With this it followed that

$$f(x) \in f(c) + F'(c + [0, 1](x - c))(x - c), \quad x \in X$$

and therefore that the inclusion

$$\tilde{f}(X) \subseteq f(c) + F'(c + (X - c)[0, 1])(X - c)$$

was valid.

This inclusion therefore requires the determination of an interval including the range of the derivative. Moore (1966) claimed that this could be accomplished using techniques for the automatic differentiation of symbolic expressions (see Chapter 11 of Moore, 1966). Based on these observations he claimed that the mean-value form was computationally simpler than the centred form. Numerical examples given in Moore (1966) showed that the mean-value form also gave better results than the standard centred form in some cases.

Both Alefeld–Herzberger (1974, 1983), Skelboe (1974) and Caprani–Madsen (1980) discussed the convergence of the mean-value form in the sense of Section 3.3. This convergence was, in particular, shown to be conditional upon the quality of the inclusion F' for f'. In Caprani–Madsen (1980) it was shown that the mean-value form was inclusion isotone. Raith (1980) discussed the forms obtained by using the arguments of Moore (1966) on the Taylor-series of a function using inclusions for the remainder term, see also Rall (1983).

Although it would seem that the mean-value form is a better choice for computing the range of a real function than the standard centred forms since it is applicable to a larger class of functions it has several disadvantages.

The main disadvantage of the mean-value form is that the problem of estimating the range of one function only has been reduced to the problem of estimating the range of another function, namely the derivative of the function. If the estimation of the derivative is too 'coarse' then the mean-value form will no longer possess the property of quadratic convergence. This problem may be alleviated by using the higher order Taylor-forms as will be shown later in this section.

An advantage is gained if the evaluation of the range of the derivative of the function is in some sense 'easier' than the evaluation of the function. This may happen for example in differential equations.

Let for example $f, g: R \to R$ and let $f'(x) = g(x)$. Then the range of f over $X \in I$ may be found using the mean-value form on

$$f(x) = \int_0^x g(t) \, dt + \text{const}$$

if the function $g(x)$ has an inclusion.

In this section we will show the mean-value form is a particular case of the general definition of a centred form given in Section 3.2 that was first discovered by Krawczyk–Nickel (1982). We will do this by defining a general Taylor-form which includes the mean-value form. Having thus related the Taylor-forms (and hence the mean-value form) to the general centred form it follows that all the properties of centred forms discussed in Sections 3.1 and 3.4 are valid for the Taylor-forms.

We first describe the Taylor-form of order k for a real function f of one variable $f: D \subseteq R \to R$. The reason for not describing the mean-value form in detail is that it is obtained as the Taylor-form of order 1. We furthermore discuss the case of a function of one variable separately since the main ideas of the Taylor-forms are contained in this case and since these forms provide an important example of centred forms. The proof for Taylor-forms for several variables which is given in the latter part of this section is obscured by the usual notational burden present in higher dimensional spaces.

For the case of one real variable we assume that $f^{(k)}$ exists on X and that $X \in I(D)$. Furthermore, it is assumed that there exists an interval estimate $F^{(k)}: I(X) \to I(R)$ of $f^{(k)}$, that is $f^{(k)}(x) \in F^{(k)}(Y)$ for all $Y \in I(X)$ and $x \in Y$. Let $\alpha: I(X) \to X$ be a developing point function in the sense of Definition 3.1.

Definition 3.3 The *Taylor-form (function)* of order k for a function f of one variable is defined by

$$T_k(Y) = f(c) + \sum_{\lambda=1}^{k-1} \frac{f^{(\lambda)}(c)}{\lambda!} (Y - c)^\lambda + \frac{F^{(k)}(Y)}{k!} (Y - c)^k \qquad (3.19)$$

for $Y \in I(X)$ and where $c = m(Y)$.

It will now be shown that this definition of a general Taylor-form satisfies the conditions of Definition 3.1 of a centred form.

We first define

$$s(x, c) = \sum_{\lambda=1}^{k-1} \frac{f^{(\lambda)}(c)}{\lambda!} h^\lambda + \frac{f^{(k)}(\theta)}{k!} h^k \qquad \text{for } x, c \in X$$

with $h = x - c$ and where the parameter θ is a function of c and x satisfying $\theta \in x \vee c$. We now seek an interval function $S: I(X) \to I$ as well as a function $G = G^1: I(X) \to I$ such that (3.1) is valid. For this let

$$S(Y) = \sum_{\lambda=1}^{k-1} \frac{f^{(\lambda)}(\alpha(Y))}{\lambda!} (Y - \alpha(Y))^\lambda + \frac{F^{(k)}(Y)}{k!} (Y - \alpha(Y))^k$$

for $Y \in I(X)$. It clearly follows using Taylor's theorem that

$$s(x, \alpha(Y)) \in S(Y) \qquad \text{for all } Y \in I(X), x \in Y.$$

Now let

$$G(Y) = \sum_{\lambda=1}^{k-1} \frac{f^{(\lambda)}(\alpha(Y))}{\lambda!} (Y - \alpha(Y))^{\lambda-1} + \frac{F^{(k)}(Y)}{k!} (Y - \alpha(Y))^{k-1} \qquad (3.20)$$

where the summation is understood to be empty for $k = 1$. Then clearly

$$S(Y) \subseteq (Y - \alpha(Y)) G(Y)$$

using (3.16) and (3.1) is satisfied. The interval

$$T_k(Y) = f(\alpha(Y)) + S(Y)$$

is called the *Taylor-form* of f on Y of order k.

The conditions for Theorem 3.2 showing the quadratic convergence of the general centred form required that the function G satisfies a Lipschitz condition as in Definition 1.3. For the Taylor-forms we now have the following theorem concerning quadratical convergence.

Theorem 3.3 *Let $f: D \subseteq R \to R$ and let $F^{(k)}$ be as defined above for $k \geq 1$. Then*
   (1) *the Taylor-form of order k is quadratically convergent if $k > 1$ provided $F^{(k)}$ is bounded,*
   (2) *the Taylor-form of order 1 (i.e. the mean-value form) is quadratically convergent if the estimation $F^{(1)} = F'$ satisfies a Lipschitz condition.*

Proof Let G be the function (3.20). For all $k > 1$ the estimation G satisfies a Lipschitz condition since it has the form

$$G(Y) = \sum_{\lambda=1}^{k-1} \frac{f^{(\lambda)}(c)}{\lambda!} (Y - c)^{\lambda - 1} + \frac{F^{(k)}(Y)}{k!} (Y - c)^{k-1}$$

which satisfies a Lipschitz condition because of Theorem 1.2 since $F^{(k)}$ is bounded.

If $k = 1$ then we have

$$G(Y) = F'(Y)$$

which satisfies a Lipschitz condition iff $F'(Y)$ does.

The result now follows from Theorem 3.2 where the Lipschitz condition on G implies quadratic convergence. □

Remark 3.4 The mean-value form, obtained by setting $k = 1$ in Definition 3.3 is clearly inferior to the higher-order Taylor-forms in the following sense: In Theorem 3.3 it is required that the estimation F' of f' satisfies a Lipschitz condition in order that the Taylor-form of order 1 (the mean-value form) is quadratically convergent. The higher order Taylor-forms ($k > 1$) only require that $F^{(k)}$ is a bounded estimate for $f^{(k)}$. In practice it is therefore much easier to work with Taylor-forms of order 2 (or higher order) since there is no explicit requirement for a Lipschitz condition.

Example 3.1 We now calculate the range of the polynomial $f(x) = x - x^2$ over the interval $X = [½ - \varepsilon, ½ + \varepsilon]$ (from Moore, 1966). We obtain $f'(x) = 1 - 2x$ and we therefore choose an estimation $F'(X)$ as

$$F'(X) = 1 - [½ - \varepsilon, ½ + \varepsilon]$$

in a natural manner. This estimate is exact. With $c = \frac{1}{2}$ we obtain the mean-value form

$$T_1(X) = f(c) + S(X) = \frac{1}{4} + \{1 - 2[\frac{1}{2} - \varepsilon, \frac{1}{2} + \varepsilon]\}[-\varepsilon, \varepsilon]$$
$$= [\frac{1}{4} - 2\varepsilon^2, \frac{1}{4} + 2\varepsilon^2].$$

The width of this mean-value form is $4\varepsilon^2$.
The standard centred form is

$$F(X) = f(c) + f'(c)(X - c) + \frac{f'(c)}{2}(X - c)^2$$
$$= \frac{1}{4} - \{[\frac{1}{2} - \varepsilon, \frac{1}{2} + \varepsilon] - \frac{1}{2}\}^2 = [\frac{1}{4} - \varepsilon^2, \frac{1}{4} + \varepsilon^2],$$

and its width is $2\varepsilon^2$.
The Taylor-form of second order is identical to the standard centred form if

$$F^{(2)}(X) = f''(c).$$

We note that although $F'(X)$ is estimated exactly, the mean-value form is a poorer estimate than the standard centred form.

We now turn to Taylor-forms for functions of several variables. Before we proceed with the development of these forms we alert the reader to the strong connections between these forms and both the Taylor-forms of one variable as well as the standard centred form in several variables. These connections will become apparent as we develop the forms.

We use the notations introduced in Section 2.5 including in particular the use of multi-indices. Therefore let $f: D \subseteq R^m \to R$ and assume that the derivative of order k exists on $X \in I^m$ for some $k \geq 1$. Assume further that each partial derivative $D^\lambda f$, $|\lambda| = k$ has an interval estimation $F^{(\lambda)}$ for which $D^\lambda f(x) \in F^{(\lambda)}(Y) \in I$ for $Y \in I(X)$ and all $x \in Y$.

Analogous to the standard centred form we use the notation $\phi_i(\lambda)$ to denote the condition

$$\lambda_1 = \ldots = \lambda_{i-1} = 0, \lambda_i > 0 \quad (i = 1, 2, \ldots, m)$$

(see also Section 3.2). Furthermore, let α again be the midpoint function. With this we may define the Taylor-form function T_k of order k as follows.

Definition 3.4 The *Taylor-form (function)* of order $k \geq 1$ on X for the real function f of m variables is defined by

$$T_k(Y) = f(c) + \sum_{|\lambda|=1}^{k-1} \frac{D^\lambda f(c)}{\lambda!}(Y - c)^\lambda$$
$$+ \sum_{|\lambda|=k} \frac{F^{(\lambda)}(Y)}{\lambda!}(Y - c)^\lambda \quad (3.21)$$

for $Y \in I(X)$ and where $c = m(Y)$.

We first show that this definition of a Taylor-form provides an inclusion function for $\tilde{f}$.

Lemma 3.5 *Using the notation of Definition 3.3 it follows that*

$$\tilde{f}(Y) \subseteq T_k(Y)$$

for all $Y \in I(X)$.

Proof Taylor's theorem in m variables states that if the derivative of order k exists then it follows that

$$f(x) = f(c) + \sum_{|\lambda|=1}^{k-1} \frac{D^\lambda f(c)}{\lambda!} (x-c)^\lambda + \sum_{|\lambda|=k} \frac{D^\lambda f(\theta)}{\lambda!} (x-c)^\lambda$$

where $\theta = \theta(x, c) \in x \vee c$. Since $F^{(\lambda)}(Y)$ is an inclusion for $\tilde{f}^{(\lambda)}(Y)$ for each $Y \in I(X)$ and $|\lambda| = k$ it follows that

$$f^{(\lambda)}(\theta) \in \tilde{f}^{(\lambda)}(Y) \subseteq F^{(\lambda)}(Y).$$

Therefore, for each $x \in Y$ we get

$$f(x) \in f(c) + \sum_{|\lambda|=1}^{k-1} \frac{D^\lambda f(c)}{\lambda!} (Y-c)^\lambda + \sum_{|\lambda|=k} \frac{F^{(\lambda)}(Y)}{\lambda!} (Y-c)^\lambda$$

which shows that $\tilde{f}(Y) \subseteq T_k(Y)$. □

We then verify that this definition of a general Taylor form indeed results in a centred form.

Lemma 3.6 *The Taylor-forms in m variables of order $k \geq 1$ are centred forms in the sense of Definition 3.1.*

Proof We again use Taylor's theorem in m variables and we write

$$s(x, c) = f(x) - f(c) = \sum_{|\lambda|=1}^{k-1} \frac{D^\lambda f(c)}{\lambda!} (x-c)^\lambda$$

$$+ \sum_{|\lambda|=k} \frac{D^\lambda f(\theta)}{\lambda!} (x-c)^\lambda$$

where $\theta = \theta(x, c)$ such that $\theta \in x \vee c$. We now seek an interval function $S: I(X) \to I$ as well as a function $G = G^1: I(X) \to I^m$ such that (3.1) is satisfied.

In order to do this we define $G(Y) = (G_1, \ldots, G_m)(Y) \in I^m$ by

$$G_i(Y) = \sum_{\substack{|\lambda|=1 \\ \phi_i(\lambda)}}^{k-1} \frac{D^\lambda f(c)}{\lambda!} (Y-c)^{\lambda-e_i} + \sum_{\substack{|\lambda|=k \\ \phi_i(\lambda)}} \frac{F^{(\lambda)}(Y)}{\lambda!} (Y-c)^{\lambda-e_i}. \qquad (3.22)$$

Since $Y - c$ is symmetric it clearly follows that

$$S(Y) = G(Y) \cdot (Y - c).$$

In the same manner as in the previous lemma it follows that $s(x, c) \in S(Y)$. The conditions of Definition 3.1 are therefore satisfied. □

Analogously to the one-dimensional case we have the following important theorem.

Theorem 3.4 *Let* $f: D \subseteq R^m \to R$ *and* $F^{(\lambda)}$ *for* $|\lambda| = k$ *be as defined above. Then*
(1) *the Taylor-form of order k is quadratically convergent if $k > 1$ provided the $F^{(\lambda)}$, $|\lambda| = k$ are bounded,*
(2) *the Taylor-form of order 1 (the mean-value form) is quadratically convergent if the $F^{(\lambda)}(Y)$, $|\lambda| = 1$ satisfy a Lipschitz condition.*

Proof Let G be as defined by (3.22). If $k > 1$ then G is Lipschitz from Theorem 1.2 and since the $F^{(\lambda)}$ for $|\lambda| = k$ are bounded. If $k = 1$ then G is Lipschitz from Theorem 1.2 since the $F^{(\lambda)}$ for $|\lambda| = 1$ are Lipschitz.

The result now follows from Theorem 3.2 where the Lipschitz condition for G implies quadratic convergence. □

Remark 3.5 The Taylor-forms may be generalized to functions on spaces more general than R^m. An example of this is given in Caprani–Madsen (1980) where a mean-value form is applied to an operator mapping $C[a, b]$ into itself.

Example 3.2 Let $f(x_1, x_2) = x_1(1 - x_1 + x_2) - x_2^2$ and let $X = ([0, 1], [0, 1])$ (see also Moore, 1976). We wish to find an estimate for the range $\bar{f}(X)$ using the mean-value form. We first obtain

$$\frac{\partial f}{\partial x_1} = 1 - 2x_1 + x_2 \quad \text{and} \quad \frac{\partial f}{\partial x_2} = x_1 - 2x_2$$

and their natural interval extensions. From this it follows that we may choose

$$G(X) = (1 - 2X_1 + X_2, X_1 - 2X_2).$$

The mean-value form for f on X is therefore

$$\begin{aligned}T_1(X) &= f(c) + (1 - 2X_1 + X_2)(X_1 - c_1) + (X_1 - 2X_2)(X_2 - c_2) \\ &= \tfrac{1}{4} + [-1, 2][-\tfrac{1}{2}, \tfrac{1}{2}] + [-2, 1][-\tfrac{1}{2}, \tfrac{1}{2}] = [-\tfrac{7}{4}, \tfrac{9}{4}].\end{aligned}$$

The range is $\bar{f}(\{[0, 1], [0, 1]\}) = [-1, \tfrac{1}{3}]$ for comparison.

3.6 DISCUSSION OF THE GENERAL DEFINITION

The definition of the general centred form is not very transparent. The following remarks and examples are given to facilitate the understanding of Definition 3.1. The rather involved nature arises from the requirement that the definition should cover as many known centred forms as possible and provide a framework for the investigation of quadratic convergence.

First, we discuss the relationship between the functions S and G in Definition 3.1. That is, we try to make it plausible that a definition of a centred form by

$$f(c) + H \cdot G(X)$$

that is, $r = 1$ and $S(X) = H \cdot G(X)$ where $H = X - c$, as is usually found does not cover sufficient examples of centred forms. We will now use Definition 3.2 instead of Definition 3.1 since convergence considerations do not play a role.

Let us look at the standard centred form calculated for a polynomial p of degree 2 using extended power evaluation as in Section 2.4. If X is any proper interval then

$$\bar{P}(X) = p(c) + p'(c)H + p''(c)\bar{H}^2/2$$

where $c = m(X)$ and $H = X - c$. Since $\bar{H}^2$ is not symmetric the same is true for

$$p'(c)H + p''(c)\bar{H}^2/2. \tag{3.23}$$

Therefore, it is never possible to represent the expression (3.23) in the form HG since the symmetry of H implies the symmetry of HG (see Section 2.4). If we set

$$S(X) = p'(c)H + p''(c)\bar{H}^2/2$$
$$\subseteq p'(c)H + p''(c)H^2/2 = HG(X)$$

where $G = p'(c) + p''(c)H/2$ then $\bar{P}(X)$ is a centred form according to our definition but not according to definitions using the form $S = HG$. Therefore, it is necessary to distinguish between S and a representation of the form $H \cdot G$.

The reason for using a *sum* $\sum_{\rho=1}^{p} H \cdot G^\rho$ instead of $H \cdot G$ as outer estimation of S is that the distributive law is not valid in interval arithmetic. This means that there are centred forms which are supersets of the corresponding inner product, $H \cdot G$ such that $S \subseteq H \cdot G$ does not hold. If, however, G is rearranged as a suitable sum,

$$G = G^1 + \ldots + G^r,$$

then

$$S \subseteq H \cdot G^1 + \ldots + H \cdot G^r.$$

We will give an example to illustrate these assertions again using the simpler Definition 3.2.

Sec. 3.6] Discussion of the General Definition

Let $f = p/q$ be a rational function in two variables where the maximum degree of the polynomials p and q is one. Then the standard centred form of order 2 on the interval $X = (X_1, X_2)$ with developing point $c = (c_1, c_2)$ and $c_1 \neq m(X_1)$, $c_2 \neq m(X_2)$, is of the form

$$F_2(X) = a_0 + a_1 H_1 + a_2 H_2 + (a_3 H_1^2 + a_4 H_1 H_2 + a_5 H_2^2)/Q(X)$$

where $a_1,\ldots,a_5 \in R$ and $H = (H_1, H_2) = X - c$. The denominator $Q(X)$ is the standard centred form of q on X (where the developing point c is as defined above).

In order to achieve a representation of the form $H \cdot G$, we must apply the denominator $Q(X)$ to each summand of the numerator. In this manner $F_2(X)$ is enlarged due to the subdistributive law and we get

$$F_2(X) \subseteq a_0 + a_1 H_1 + a_2 H_2 + A_3 H_1^2 + A_4 H_1 H_2 + A_5 H_2^2 \qquad (3.24)$$

where $A_i = a_i/Q(X)$. It is now easy to rearrange the right side of (3.24) as an inner product $H \cdot G$. The intervals H_i are, however, not symmetric which implies that the distributive law cannot be applied and the subdistributive law is valid then in the direction opposite to that of (3.24). We therefore obtain

$$a_0 + a_1 H_1 + \ldots + A_5 H_2^2 \supseteq a_0 + H_1 G_1 + H_2 G_2 = a_0 + H \cdot G \qquad (3.25)$$

where for instance,

$$G_1 = a_1 + (a_3 H_1 + a_4 H_2)/Q(X),$$
$$G_2 = a_2 + A_5 H_2.$$

Because of the different directions of the inclusions in (3.24) and (3.25) it is not possible to compare $F_2(X)$ and $a_0 + H \cdot G$ directly. If, however, more than one function G is admitted, then one can proceed as follows. Depending on the form of the expression for $F_2(X)$, functions $G^\rho = (G_1^\rho, G_2^\rho)$ can be defined in a natural manner,

$$G^1 = (a_1, a_2),$$
$$G^2 = (A_3 H_1, 0),$$
$$G^3 = (A_4 H_2, 0),$$
$$G^4 = (0, A_5 H_2),$$

such that

$$F_2(X) = f(c) + S$$

with $a_0 = f(c)$ and such that

$$S \subseteq H \cdot G^1 + \ldots + H \cdot G^4$$

is valid. If numerical data is inserted then it can be seen that these considerations are not only of a theoretic nature.

The second part of our discussion deals with the significance of the function α with respect to quadratic convergence. In a heuristic sense we may explain the quadratic convergence in the following manner. Suppose we have computed a centred form estimate $F(X)$ for $\tilde{f}(X)$ such that $\tilde{f}(X) \subseteq F(X)$. If we now wish to improve the estimate $F(X)$ by subdividing X into two (or more) subintervals Y_1 and Y_2 such that $X = Y_1 \cup Y_2$ and calculating the centred forms $F(Y_1)$ and $F(Y_2)$ then we obtain

$$\tilde{f}(X) = \tilde{f}(Y_1) \cup \tilde{f}(Y_2) \subseteq F(Y_1) \cup F(Y_2)$$

which is an improvement on $F(X)$ if the centred form function F is inclusion isotone. The partioning is now applied to Y_1 and Y_2, and so on, and the generated sequence of inclusions converges quadratically to $\tilde{f}(X)$ if the Lipschitz condition for the corresponding function G holds. The method itself is called the subdivision method and it is extensively discussed in Chapter IV. These few lines show, however, that we must be able to reduce the given domain X to subintervals $Y \in I(X)$ and to choose a corresponding developing point c for each Y in the subdivision. This correspondence is governed by the function α, and D_0 is provided as the set or the superset of all developing points that occur in an actual subdivision or in a special class of subdivisions.

Some of the definitions found in the literature do not take this connection between Y and the developing point $\alpha(Y)$ into account. They use a fixed c such that it is not possible to use the definition with respect to the subdivision method. There is no possible way of investigating the quadratic convergence without considering this connection.

Further, the developing point function α plays an important role if the centred form definition is used in the description of iterative methods, see Krawczyk (1982) and Section 6.6. As an example, let us consider the Krawczyk iteration operator (Krawczyk, 1969). Let X be the starting interval containing a zero x_0 of a function f where x_0 is to be located, F' an interval function including the derivative f', that is

$$f'(x) \in F'(Y) \quad \text{for all } Y \in I(X) \text{ and } x \in Y,$$

and assume $0 \notin F'(X)$ which implies that x_0 is the only zero of f in X. If the interval sequence $(Y_n)_{n=0}^{\infty}$ is defined by $Y_0 = X$ with $(z_n)_{n=0}^{\infty}$ being a sequence such that $z_n \in R$, $z_n \neq 0$ and if

$$Y_{n+1} = \left[\alpha(Y_n) - \frac{f(\alpha(Y_n))}{z_n} + \left(1 - \frac{F'(Y_n)}{z_n}\right)(Y_n - \alpha(Y_n)) \right] \cap Y_n,$$

then the condition $\alpha(Y_n) \in Y_n$ is already sufficient for (Y_n) to be a nested sequence, that is, $Y_0 \supseteq Y_1 \supseteq \ldots$ and to converge to the zero x_0. Usually, $\alpha(Y_n)$ is again chosen as midpoint of Y_n.

3.7 PRACTICAL CONSIDERATIONS FOR CONSTRUCTING CENTRED FORMS

In this section we discuss when it makes sense to construct centred forms as well as some practical guidelines for their actual construction. We do not consider the requirement of a user who is only concerned with the computation of any inclusion (arbitrarily poor) for the range of a given function f. The effort of computing a centred form would not be warranted for such a user. A simple inclusion may almost always be found by partitioning the expression for f into appropriate subexpressions such that inclusions may easily be found for the subexpressions. Interval arithmetic may then be used to compute an inclusion for the whole expression. The following two examples will illustrate this technique.

Example 3.3 Let

$$f(x, y, z) = \frac{x^{1/2}[\sin y + e^z/(9 + x + y - 3z)] + xyz/\log(1 + x)}{\cos(xy)\sin z + xy}$$

and $x, y, z \in X = [2, 4]$ let also $Y = X \times X \times X$. If we wish to find any inclusion for the range $\tilde{f}(Y)$ then we can proceed as follows: First, inclusions for the subexpressions are developed as follows:

$x^{1/2} \in [1, 2]$, $\sin y \in [-1, 1]$, $e^z \in [e^2, e^4] \subseteq [2^2, 3^4] = [4, 81]$,

$x + y - 3z \in [2, 4] + [2, 4] - 3[2, 4] = [-8, 2]$, $xyz \in [2, 4]^3 = [8, 64]$,

$\log(1 + x) \in [\log 3, \log 5] \subseteq [\log e, \log 2e] \subseteq [1, \log 2 + \log e]$

$\subseteq [1, \log e + \log e] = [1, 2]$, $1/\log(1 + x) \in [½, 2]$,

$\cos(xy) \sin z \in [-1, 1]$, $xy \in [4, 16]$.

Secondly, the subexpressions are replaced by their inclusions and interval arithmetic is used to obtain an inclusion of the range using also Lemma 1.4,

$$\tilde{f}(Y) \subseteq \frac{[1, 2]([-1, 1] + [4, 81]/[1, 11]) + [8, 64][½, 1]}{[-1, 1] + [4, 16]}$$

$\subseteq [0, 76]$.

The next example shows that discontinuous function may also be handled by this method.

Example 3.4 Let $[x]$ be the largest integer which is not larger than a given integer number x, for instance, $[3] = 3$, $[-7] = -7$, $[3.3] = 3$, $[-3.3] = -4$. Let

$$f(x, y, z) = (x - [x])(\text{sgn}(2 - y))/[z]$$

for $x, y, z \in X = [1, 4]$, $Y = X \times X \times X$. Then

$$x - [x] \in [0, 1], \text{sgn}(3 - y) \in [-1, 1], [z] \in [1, 4],$$

and

$$\bar{f}(Y) \subseteq [0, 1][-1, 1]/[1, 4] = [-1, 1].$$

Although a user may be only interested in a roughly estimated inclusion of the range, it is sometimes necessary to improve this estimation. This may for example happen if one wishes to check whether $0 \notin \bar{f}(X)$ or not in cases where one wishes to assure that f has no zeros. If $S \supseteq \bar{f}(X)$ is an estimation with $0 \notin S$ then one can be certain that $0 \notin \bar{f}(X)$. If S is estimated too roughly then frequently $0 \in S$ will occur. This means that $0 \in \bar{f}(X)$ and $0 \notin \bar{f}(X)$ is possible. In order to obtain a decision it is necessary to improve S, for instance, by a sequence of inclusions,

$$S \supseteq \ldots \supseteq S_n \supseteq S_{n+1} \supseteq \ldots \supseteq \bar{f}(X).$$

If there exists an n such that $0 \notin S_n$, the question is solved. If there does not exist such n, then the question is theoretically solved if S_n converges to $\bar{f}(X)$. The method may, however, not be realized by a computer program because of the finite representation of the reals by computers.

In general it is not sensible to construct centred forms for functions of the type shown in Example 3.4 that are not continuously differentiable. The reason is that the advantages of centred forms are based on an analytic (and not only an arithmetic) rearrangement of the difference $f(x) - f(c)$ which is only possible if proper differentiability conditions exist such that the factor $(x - c)$ may be split off from the difference with the remainder being at least bounded. For instance, the standard centred form depends on the Taylor expansion of f. If $f(x) - f(c)$ can only be treated arithmetically, that is, only using the four arithmetic operations without using limit operations, then, a general improvement of the range estimation cannot be guaranteed or observed in practice. In order to make this discussion plausible we give a simple but typical example.

Example 3.5 Let $f(x) = x - [x]$ be defined on the interval $X = [0, 2]$. Clearly, $\bar{f}(X) = [0, 1]$. If we proceed automatically in order to get an inclusion of the range, then we get as 'natural' interval extension,

$$f(X) = X - [X]$$

where $[X]$ is the smallest interval that contains the set $\{[x]: x \in X\}$, and

Sec. 3.7]   Practical Considerations for Constructing Centred Forms   87

$$f(X) = [0, 2] - [0, 2] = [-2, 2].$$

If the real representation is used that underlies the centred form, see Definition 3.1, it is still possible to determine the function s,

$$s(x, c) = f(x) - f(c) = x - [x] - c + [c]$$
$$= (x - c) - ([x] - [c]).$$

Treating the term $[x] - [c]$ is more complicated than treating $x - [x]$, and this rearrangement furthermore does not gain any advantage. If we try to split off the factor $(x - c)$ and to represent s in the form

$$s(x, c) = (x - c) g(x, c),$$

as in (2.2) and to find a reasonable inclusion for g, then this procedure fails, since on the one hand, the function

$$g(x, c) = 1 - ([x] - [c])/(x - c)$$

is unbounded and on the other, it contains a zero in each interval extension of the denominator which means that the division is not allowed.

We will now give some directions that may be helpful in order to find a centred form function as required in Definition 3.1.

Let $X \in I^m$ and let $f: X \to R$. The following steps are generally successful in constructing an inclusion function for $\bar{f}$:

1. Find a function $g: X \times X \to R^m$ such that $s(x, c) = f(x) - f(c) = g(x, c) \cdot (x - c)$ for all $x, c \in X$.
2. Find an interval function $G: I(X) \to I^m$ such that $g(x, c) \in G(Y)$ for all $Y \in I(X), x \in Y, c = m(Y)$.
3. The function $F: I(X) \to I$. defined by $F(Y) = f(c) + G(Y) \cdot (Y - c)$ where $c = m(Y)$ is a centred form function for f on X.
4. In order to get reasonable and natural representations and expressions it is frequently necessary to split the functions g and G,

$$g = \sum_{\rho=1}^{r} g^\rho \quad \text{and} \quad G = \sum_{\rho=1}^{r} G^\rho$$

such that $g^\rho(x, c) \in G^\rho(Y)$ for all $Y \in I(X), x \in Y, c = m(Y)$, see Section 3.6.

5. In order to obtain improvements (distributive law, extended power evaluation) one may choose a function $S: I(X) \to I$ such that

$$S(Y) \subseteq \sum_{\rho=1}^{r} G^\rho \cdot (Y - c) \quad \text{for all } Y \in I(X), c = m(Y).$$

Having done this, one has to check that the improvement S is not too good, that is, S still has to satisfy the condition

$$s(x, c) \in S(Y) \quad \text{for all } Y \in I(X), x \in Y, c = m(Y).$$

6. In order to get a quadratically convergent centred form function one has to make certain that the necessary assumptions for G are valid, namely that G satisfies a Lipschitz condition as seen in Section 3.3.

If the reader is less interested in the sophisticated considerations treated in this section so far, but more interested in an easy recipe for getting a quadratically convergent centred form function that nearly always works, then an approximation of the Taylor-form function of second order (and not of the mean-value form) is highly recommended.

Recipe 1 If f is a rational function in one or several variables the standard centred or Krawczyk's form function yields quadratically convergent inclusions of the ranges $\bar{f}(Y)$.

Recipe 2 Assume that $f: X \to R$ is twice differentiable on the interval $X \in I^m$. Fore each multi-index $\lambda \in N^m$ with $|\lambda| = 2$ find a bounded interval function $F^\lambda: I(X) \to I$ such that

$$D^\lambda f(x) \in F^\lambda(Y) \quad \text{for all } x \in Y, Y \in I(X).$$

Then the function $F: I(X) \to I$ defined by

$$F(Y) = f(c) + \sum_{|\lambda|=1} D^\lambda f(c) (Y - c)^\lambda + \sum_{|\lambda|=2} F^\lambda(Y) (Y - c)^\lambda/\lambda!$$

where c is an abbreviation of $m(Y)$, is a centred form function for f on X and quadratically convergent, as seen in Section 3.5.

The proof of the assertions of the recipes are given in 3.2 and 3.5.

It is very convenient to use Recipe 2 since the functions F^λ are only required to be *any* (arbitrarily bad) inclusion isotone estimation which can be as simple as a constant interval, for instance

$$F^\lambda(Y) = [a, b] \in I \quad \text{for all } Y \in I(X) \text{ and all } \lambda \text{ with } |\lambda| = 2.$$

Therefore we need only bounds of the partial derivatives $D^\lambda f$ for $|\lambda| = 2$, and as discussed in Section 3.5 it might seem that Recipe 2 only provides a shift of the task of finding boundaries from the function f to the partial second derivatives. We repeat that this conjecture is wrong because we have mentioned at the beginning of this section that *any* inclusion nearly always can be found and we will now focus on the search for *arbitrarily good* estimations. The recipe says that if *any* (and only one set of) bounded inclusions for the second partial derivatives are found then the quadratic convergence of F, see Theorem 3.4 is guaranteed.

If we compare Recipe 2 with the mean-value form function for f we notice that for the latter method only differentiability is assumed for f but that there are some further conditions which are not too easy to realize. If f is therefore twice differentiable we recommend Recipe 2 in any case, if

Sec. 3.7] Practical Considerations for Constructing Centred Forms

not, then clearly the mean-value form functions still provide an interesting possibility.

Let us give some simple examples for Recipe 2.

Example 3.6 Let $f(x) = x \cos x$ for $x \in X = [-\pi, \pi]/2$. We get $f'(x) = \cos x - x \sin x$ and $f''(x) = -2 \sin x + x \cos x$ where $f''(x) \in [-2, 2] + ([-\pi, \pi]/2) [-1, 1] \subseteq [-6, 6]$. Using the recipe, the interval function $F: I(X) \to I$ given by

$$F(Y) = f(c) + f'(c)(Y - c) + [-3, 3](Y - c)^2 \quad \text{for } Y \in I(X)$$

where c is written instead of $m(Y)$ is a quadratically convergent centred form function for f. The reader may apply the mean-value form and compare the effort.

In the next example we consider a monotone function in order to be able to calculate the width $w[\tilde{f}(Y)]$ directly and to compare it with the width of the centred form function.

Example 3.7 Let $f(x) = e^x$ for $x \in X = [0, 1]$. We get $f'(x) = e^x$ and $f''(x) = e^x \in [0, 4]$. By Recipe 2, a quadratically convergent centred form function for f is given by

$$F(Y) = e^c + e^c(Y - c) + [0, 2](Y - c)^2 \quad \text{for } Y \in I(X)$$

where c is an abbreviation of $m(Y)$. We shall now verify the quadratic convergence directly by using Taylor's formula and equation (1.3):

$$w[F(Y)] = e^c w(Y) + 2w(Y)^2,$$

and, if $Y = [y, z]$,

$$\begin{aligned} w[\tilde{f}(Y)] &= e^z - e^y = e^z - e^c + e^c - e^y \\ &= e^c(z - c) + e^\xi(z - c)^2/2 - e^c(y - c) - e^\eta(y - c)^2/2 \\ &= e^c w(Y) + (e^\xi - e^\eta) w(Y)^2/8, \end{aligned}$$

where $\xi \in [c, z]$ and $\eta \in [y, c]$. Therefore

$$w[F(Y)] - w[\tilde{f}(Y)] = 2w(Y)^2 + (e^\eta - e^\xi) w(Y)^2/8$$
$$\leqslant 2w(Y)^2$$

for every $Y \in I(X)$.

3.8 THE KNOWLEDGE ABOUT MONOTONICITY

If a function f is monotone in some of the variables then the complexity of computing the centred form of f is reduced. If f is a continuous function in one variable, the case is evident, and we have

$$\bar{f}([a, b)] = f(a) \vee f(b)$$

if defined. If a function $f(x_1,\ldots,x_m)$ is defined for $x_i \in X_i \in I (i = 1,\ldots,m)$, we say that f is monotone in the variable x_k if for each choice of values $c_i \in X_i$ ($i = 1,\ldots,k-1, k+1,\ldots,m$), the function

$$g(x_k) = f(c_1,\ldots,c_{k-1}, x_k, c_{k+1},\ldots,c_m)$$

is monotone. The facilities of using monotonicity properties are based on the following lemma which is due to Skelboe (1974).

Lemma 3.7 *Let the continuous function $f(x_1,\ldots,x_m)$ be defined for $x_i \in X_i$ ($i = 1,\ldots,m$) and monotone in x_1 (without restricting the generality). If $X_1 = [a, b]$ and the function g_c is defined for any $c \in X_1$ by $g_c(x_2,\ldots,x_m) = f(c, x_2,\ldots,x_m)$ for $x_i \in X_i$ ($i = 2,\ldots,m$), then*

$$\bar{f}(X_1,\ldots,X_m) = \bar{g}_a(X_2,\ldots,X_m) \vee \bar{g}_b(X_2,\ldots,X_m).$$

Proof Assume $y = f(c_1,\ldots,c_m)$ for some $c_i \in X_i$ ($i = 1,\ldots,m$). Since $a \leq c_1 \leq b$ using the monotonicity we get

$$g_a(c_2,\ldots,c_m) \leq f(c_1,\ldots,c_m) \leq g_b(c_2,\ldots,c_m)$$

or the opposite chain, and the value y lies in the interval hulls $g_a(c_2,\ldots,c_m) \vee g_b(c_2,\ldots,c_m) \subseteq \bar{g}_a(X_2,\ldots,X_m) \vee \bar{g}_b(X_2,\ldots,X_m)$. The continuity of f is used for finding a term $f(c_1,\ldots,c_m)$ which lies between given values $f(a, d_2,\ldots,d_m)$ and $f(b, e_2,\ldots,e_m)$ in order to prove the inclusion in the opposite direction. □

If this lemma is applied repeatedly then it can be generalized to functions monotone in several variables.

Therefore, if it is required to determine the range of the function $e^x/(x^2 + y)$ for $x \in X = [-1, 1]$ and $y \in [3, 4]$ then it is sufficient to determine the interval hull of the (exact or estimated) ranges of the functions $g_3(x) = e^x/(x^2 + 3)$ and $g_4(x) = e^x/(x^2 + 4)$ for $x \in X$. The ranges $\bar{g}_3(X)$ and $\bar{g}_4(X)$ can be calculated by using the centred forms $G_3(X)$ and $G_4(X)$.

Corollary 3.1 *If the continuous function f is representable in the form*

$$f(x_1,\ldots,x_m) = g(x_1,\ldots,x_{m-1}) * h(x_m) \qquad (3.26)$$

for $x_i \in X_i$ ($i = 1,\ldots,m$) and $* \in \{+, -, \cdot, /\}$, where h is a monotone function, then

$$\bar{f}(X_1,\ldots,X_m) = \bar{g}(X_1,\ldots,X_{m-1}) * \bar{h}(X_m).$$

Proof Since x_m is separated, the interval hull which is provided in Lemma 3.7 has the desired form. □

There are two advantages of this corollary. The first is that it is not necessary to calculate the interval hull, the second is that an inclusion procedure such as the centred form is only required once, i.e. for the inclusion of $\bar{g}(X_1,...,X_{m-1})$.

Since each occurrence of a variable in which the function f is monotone diminishes the number of variables in the centred form, it is suggested to manipulate the expression of f in a way that allows the use of Corollary 3.1 as often as possible, cf. the following example of Skelboe (1974), Moore (1976). Let

$$f(x, y, z) = \frac{x + y}{x - y} z \quad \text{for } x \in X = [1, 2], y \in Y = [5, 10], \text{ and } z \in Z$$

$= [2, 3]$, then an optimal arrangement of f is

$$f(x, y, z) = \left(1 + \frac{2}{(x/y) - 1}\right) z$$

such that the range is determined directly by

$$\bar{f}(X, Y, Z) = \left(1 + \frac{2}{(X/Y) - 1}\right) Z.$$

If f is a function in the two variables (for simplicity) $x \in X$ and $y \in Y$ it may happen that f is monotone in y but that f is not representable in the form (3.26), for example, the function $f(x, y) = e^{(x-1)y}(x + 1)$. In such a case there seems to be no advantage in simply replacing y by Y, since, when applying the centred form, the variable y will occur several times in the centred form formula. Thus, the use of Lemma 3.7 is again recommended.

The knowledge about monotonicity is also helpful in making the subdivision method more efficient, i.e., the following lemma will be used. But first we need some notions.

If a variable x_k occurs in an arithmetic expression $f(x_1,...,x_m)$ at most once and of first order, then for brevity x_k is said to *occur only once* in this expression. Thus x_3 occurs only once in $(x_1 + x_2^2)/(3x_3)$ or in x_2^2 or in $x_2 x_3^{-1}$ (since x_3^{-1} is the abbreviation for $1/x_3$), but x_3 does not occur only once in $x_1 + 3x^3$ or in x_3/x_3.

Lemma 3.8 *Let $f(x_1,...,x_m)$ be an arithmetic expression in which each variable occurs only once. Then*

$$f(X_1,...,X_m) = \bar{f}(X_1,...,X_m) \quad \text{for } X_1,...,X_m \in I$$

if defined.

Proof One only has to show that $f(X_1,...X_m) \subseteq \bar{f}(X_1,...X_m)$. Let $y \in f(X_1,...X_m)$. Since each variable occurs only once, there exist reals $c_i \in X_i$ such that $y = f(c_1,...,c_m) \in \bar{f}(X_1,...X_m)$. □

The occurrence and distribution of monotone functions in the class of all functions is investigated by Nickel (1977).

CHAPTER IV

More about quadratic convergence

In the previous chapters we developed the centred form for computing a good inclusion for the range of a function over an interval. The inclusion was considered to be good if it converged quadratically to the range of the function when the width of the domain tended to zero. The original problem was, however, to compute a good inclusion on a fixed domain. This may be stated as: Given a function $f: D \subseteq R^m \to R$, a fixed interval $X \in I(D)$ and $\varepsilon > 0$ find an including estimate $F(X)$ for the range $\bar{f}(X)$ of f such that

$$w[F(X)] - w[\bar{f}(X)] < \varepsilon.$$

Clearly it is not possible in general to compute such an estimate using only the centred form.

We therefore develop and discuss the subdivision method due to Moore (1966) as well as refinements due to Skelboe (1974), Moore (1976, 1979) and Asaithambi–Zuhe–Moore (1982). When we now combine the centred form with the subdivision method we obtain an effective tool for computing including estimates of the range of functions of required accuracy.

This chapter is divided into three parts. We first introduce and define the subdivision method. We then show that the subdivision method is very effective when combined with an estimate such as a quadratically convergent centred form. Finally, we discuss the method given by Skelboe (1974), eliminating the logical flaws in the method.

4.1 THE SUBDIVISION METHOD

The subdivision method was originally developed by Moore (1966). It was based on the simple idea that the interval arithmetic estimate of the range of a rational function over an interval could be improved by subdividing the interval and then computing the union of the interval estimates over the subintervals for an improved estimate.

In this section we develop the *subdivision method* in the more general setting of including functions.

We first introduce some notation for the uniform subdivision of an interval $X = (X_1, \ldots, X_m) \in I^m$. Let $n \geq 0$ be a given fixed integer. For each $X_i = [x_i, y_i]$, $i = 1, 2, \ldots, m$ we define the subintervals

$$X_{ij} = [x_i + (j-1) w(X_i)/n, x_i + jw(X_i)/n],$$

$$j = 1, 2, \ldots, n.$$

Clearly $X_i = \bigcup_{j=1}^n X_{ij}$, $i = 1, 2, \ldots, m$. If we now define the index set

$$J_m = \{1, 2, \ldots, n\}^m$$

then it follows that

$$X = \bigcup_{\tau \in J_m} X_\tau$$

where

$$X_\tau = (X_{1,j_1}, \ldots, X_{m,j_m}) \quad \text{for } \tau = (j_1, j_2, \ldots, j_m) \in J_m.$$

From the definition of X_{ij} and X_τ it follows that

$$w(X_\tau) = \frac{w(X)}{n} \quad \text{for all } \tau \in J_m. \tag{4.1}$$

Let now $f: D \subseteq R^m \to R$, $X \in I(D)$ and assume that there exists a general inclusion function of $\bar{f}$ over X, that is, a function $F: I(X) \to I$ with the property that

$$\bar{f}(Y) \subseteq F(Y) \quad \text{for all } Y \in I(X). \tag{4.2}$$

We then define

$$S_n(X) = \bigcup_{\tau \in J_m} F(X_\tau) \tag{4.3}$$

to be the *n*th *uniform refinement* (or *refinement* for brevity) of F. Clearly

$$S_n(X) = \bigcup_{\tau \in J_m} F(X_\tau) \supseteq \bigcup_{\tau \in J_m} \bar{f}(X_\tau) = \bar{f}(X)$$

from the inclusion property (4.2). The *subdivision method* then simply consists of computing refinements S_n for $n = 1, 2, \ldots$ or for some subsequence.

Moore (1966) furthermore gives some estimations of the computational complexity of the subdivision method.

4.2 GLOBAL CONVERGENCE IN THE SUBDIVISION METHOD

In the previous section we introduced the subdivision method. By applying it to an inclusion function $F: I(X) \to I$ of a range function $\tilde{f}: I(X) \to I$, estimations $S_n(X) \supseteq \tilde{f}(X)$ were obtained. We will now consider the convergence properties of this method, in particular we will show that if F converges quadratically to $\tilde{f}$, then $S_n(X)$ converges quadratically to $\tilde{f}(X)$ as $n \to \infty$. In the subdivision method we therefore have a tool whereby we may compute improved estimates $S_n(X)$ of the range $\tilde{f}(X)$ over the interval X by allowing the computations to progress over successively smaller intervals X_{ij}.

Let therefore $X \in I^m$, $f: X \to R$ and $F: I(X) \to I$ be any inclusion function for the range function $\tilde{f}: I(X) \to I$ such that

$$\tilde{f}(Y) \subseteq F(Y) \quad \text{for all } Y \in I(X)$$

Let now $S_n(X)$ be the nth refinement of $F(X)$ obtained from (4.3). Then the following is valid.

Theorem 4.1 *If the inclusion function F for $\tilde{f}$ on X is quadratically convergent to the range function $\tilde{f}$, then*

$$w[S_n(X)] - w[\tilde{f}(X)] = O(1/n^2).$$

Proof Let Y_l and Y_r be the intervals of the nth subdivision of X, which determine $S_n(X)$, that is,

$$S_n(X) = F(Y_l) \vee F(Y_r)$$

where the left endpoint of $\tilde{f}(X)$ is contained in $F(Y_l)$ and the right endpoint of $\tilde{f}(X)$ is contained in $F(Y_r)$. (Studying the subdivision method one can check that such intervals Y_l and Y_r do exist.) From the assumption, it follows that

$$w[F(Y)] - w[\tilde{f}(Y)] = O(1/n^2) \quad \text{for } Y \in \{Y_l, Y_r\},$$

and from $\tilde{f}(X) \subseteq \tilde{f}(Y_l) \vee \tilde{f}(Y_r)$ we obtain

$$w[S_n(X)] - w[\tilde{f}(X)] \leq w[F(Y_l)] - w(\tilde{f}(Y_l))]$$
$$+ w[F(Y_r)] - w(\tilde{f}(Y_r))] = O(1/n^2). \qquad \square$$

Remark 4.1 If the inclusion function F for $\tilde{f}$ on X is only linearly convergent to $\tilde{f}$ then the proof of Theorem 4.1 shows that

$$w[S_n(X)] - w[\tilde{f}(X)] = O(1/n).$$

Remark 4.2 Both Theorem 4.1 and the previous remark are still valid if the subdivision is not uniform. In this case, however, there must exist a constant γ such that the subintervals X_{ij} of the nth subdivision satisfy

$$w(X_{ij}) \leq \gamma/n$$

where X_{ij} is the jth subinterval of X_i and where the constant γ is independent on n. Such a subdivision can be useful if $w(X_i)/w(X)$ is very small for some coordinate direction i in that it may reduce the number of evaluations, see also Example 4.1.

Example 4.1 A large saving in computations may be possible using the idea in Remark 4.2. Let for example the range function $\bar{f}$ of $f: D \subseteq R^m \to R$ have an inclusion function F with a Lipschitz constant L. Furthermore, let $X = (kY, Y, ..., Y) \in I(D)$ for some $k \geq 0$. Then $w(X) = kw(Y)$.

In order to compute a refinement $S_n(X)$ by the uniform subdivision method that satisfies

$$w[S_n(X)] - w[(\bar{f}(X)] \leq \varepsilon$$

for some prescribed accuracy $\varepsilon > 0$ one can use the estimation

$$w[S_n(X)] - w[\bar{f}(X)] \leq 2Lw(X)/n$$

which can either be derived from the proof of Theorem 4.1 using the constant L or can be found in Moore (1979). Thus, in order to attain the required accuracy, $n \geq 2Lw(X)/\varepsilon$ must be chosen and therefore approximately

$$\left(\frac{2Lw(X)}{\varepsilon}\right)^m$$

evaluations $F(X_{ij})$ are necessary for computing $S_n(X)$. However, if with respect to Remark 4.2 the subdivision is such that at the first step only $X_1 = kY$ is uniformly divided into k intervals of length $w(Y)$ and no subdivision of $X_2, ..., X_m$ is done, and then the further steps of the subdivision are a uniform subdivision of the generated k subintervals in all directions, then only approximately

$$\left(\frac{2Lw(X)}{\varepsilon}\right)^m / k^{m-1}$$

evaluations $F(X_{ij})$ are necessary in order to attain the accuracy ε.

In order to compute the nth uniform refinement $S_n(X)$ we have to evaluate F over each X_τ, $\tau \in J_m$. Since the set J_m contains n^m elements it is clear that even for moderate dimensions the amount of computational effort may be rather large. For 10 subdivisions in three dimensions we already get 1000 evaluations. We therefore consider some further improvements to the subdivision method that either depend on the function to be evaluated or on the results already obtained from a previous

Sec. 4.2]   Global Convergence in the Subdivision Method   97

calculation. The second and most important improvement is the recursive method of Skelboe (1974) which will be discussed in Section 4.3. Here we discuss two ways in which the subdivision method may be improved if further information about the function f is available.

The following two theorems due to Skelboe (1974) may aid in reducing the number of evaluations in the subdivision method substantially by reducing the number of coordinate directions in which a subdivision is necessary. The theorem is valid for inclusion functions that are obtained as natural interval extensions to rational functions.

Theorem 4.2 Let $D \subseteq R^m$, $X = (X_1, \ldots, X_m) \in I(D)$, and $n > 0$. Let $f: D \to R$ be a rational function and let $\bar{f}(Y)$ be the natural interval extension of $f(x)$ on $Y \in I(X)$. If the component x_k of x occurs only once in the expression for f, then

$$f(X_1, \ldots, X_k, \ldots, X_m) = \bigcup_{j=1}^{n} f(X_1, \ldots, X_{kj}, \ldots, X_m)$$

where X_{kj} is defined as in Section 4.1.

Proof If we replace the first occurrence of x_1 by $x_1^{(1)}$, the second by $x_1^{(2)}$, etc., and in general, the jth occurrence of x_i by $x_i^{(j)}$ then an expression $h(x_1^{(1)}, \ldots, x_k^{(1)}, \ldots)$ is generated in which each variable $x_i^{(j)}$ occurs exactly once. Let σ_i be the number of occurrences of x_i in f, then the expression $h(x_1^{(1)}, \ldots, x_k^{(1)}, \ldots)$ generates f if $x_i^{(j)} = x_i$ for $j = 1, \ldots, \sigma_i$.

Now let $Y^* = Y^{\sigma_1} \times \ldots \times Y^{\sigma_m} \in I^{\sigma_1 + \ldots + \sigma_m}$ be defined for any $Y \in I(X)$. From Lemma 3.8, since each variable $x_i^{(j)}$ occurs only once in h, we have

$$\bar{h}(Y^*) = h(Y^*)$$

where $h(Y^*)$ is the natural interval extension of h to Y^*. Furthermore, by the definition of a natural interval extension, we have

$$f(Y) = h(Y^*).$$

Since $X_k = \bigcup_{j=1}^{n} X_{kj}$ using the just mentioned relations we get,

$$f(X) = h(X^*) = \bar{h}(X^*)$$

$$= \bigcup_{j=1}^{n} \bar{h}(X_1^{\sigma_1}, \ldots, X_{kj}, \ldots, X_m^{\sigma_m})$$

$$= \bigcup_{j=1}^{n} h(X_1^{\sigma_1}, \ldots, X_{kj}, \ldots, X_{m_m}^{\sigma})$$

$$= \bigcup_{j=1}^{n} f(X_1, \ldots, X_{k-1}, X_{kj}, X_{k+1}, \ldots, X_m). \quad \square$$

The Theorems 4.1 and 4.2 will now be combined for the following result: if the variables $x_{k_1}, \ldots, x_{k_\rho}$ occur only once in a rational expression $f(x_1, \ldots,$

x_m), if F is, as natural interval extension of f, a quadratically convergent inclusion function to the range function $\tilde{f}$, and if the subdivision is done only for the coordinate directions $\{1,\ldots, m\}\backslash\{k_1,\ldots, k_p\}$ when using the subdivision method, then the quadratic convergence of the inclusions to the range $\tilde{f}(X)$ in the sense of Theorem 4.1 is retained. The justification for this assertion is given by the following theorem.

Theorem 4.3 *Let f be as in Theorem 4.2 such that the variable x_k occurs only once in the expression for f. If the natural interval extension $f(Y)$ of f to $Y \in I(X)$ converges quadratically to $\tilde{f}(Y)$ and if $S_n^*(X)$ is the inclusion computed by a uniform subdivision of all components of X except X_k then*

$$w[S_n^*(X)] - w[\tilde{f}(X)] = O(1/n^2).$$

Proof Let the index set $J_m = \{1,\ldots, n\}^m$, the intervals X_τ for $\tau \in J_m$, and the nth refinement be defined as in Section 4.1. To each $\tau = (j_1, \ldots, j_m) \in J_m$ let the $(m-1)$-tuple $\tau^* = (j_1, \ldots, j_{k-1}, j_{k+1}, \ldots, j_m)$ and the subinterval

$$X_{\tau^*} = (X_{1j_1}, \ldots, X_{k-1,j_{k-1}}, X_k, X_{k+1,j_{k+1}}, \ldots, X_{mj_m})$$

be defined. The X_{τ^*} are precisely those subintervals that occur in the nth refinement $S_n^*(X)$. If $\tau \in J_m$, then by Theorem 4.2 we get

$$f(X_{\tau^*}) = \bigcup_{\substack{\sigma \in J_m \\ \sigma^* = \tau^*}} f(X_\sigma),$$

and from this it follows that

$$S_n^*(X) = \bigcup_{\tau \in J_m} f(X_{\tau^*}) = \bigcup_{\tau \in J_m} \bigcup_{\substack{\alpha \in J_m \\ \sigma^* = \tau^*}} f(X_\sigma),$$

$$= \bigcup_{\tau \in J_m} f(X_\tau) = S_n(X).$$

The quadratic convergence of $S_n(X)$ and hence of $S_n^*(X)$ follows from Theorem 4.1. □

Remark 4.3 Another procedure for reducing the computational effort of computing estimates for the range $\tilde{f}(X)$ for $f: D \subseteq R^m \to R$ for some $X \in I(D)$ was suggested by Moore (1975) (see also Moore, 1979). For this we assume that the partial derivatives

$$f_i(x) = \frac{\partial f(x)}{\partial x_i}$$

exist and have inclusions

$$F_i(Y) \supseteq \tilde{f}_i(Y) \quad \text{for each } Y \in I(X).$$

If it now turns out that

$$0 \leq \min F_i(Y)$$

for some i, $1 \leq i \leq m$ and for some $Y \in I(X)$ then we may conclude that

$$g(x_i) = f(x_1, \ldots, x_i, \ldots, x_m)$$

is a monotonically increasing function of f in x_i over Y and the techniques of Section 3.7 may be used for Y. A similar statement is valid if

$$0 \geq \max F_i(Y).$$

Further, various hints to reduce the number of function evaluations at the subdivision method are given by Moore (1976, 1979), Asaihambi–Zuhe–Moore (1982), and Hansen (1980).

4.3 THE METHOD OF SKELBOE

The subdivision method discussed in Section 4.1 has a very large computational complexity. In the method suggested by Skelboe (1974) the number of computations required is reduced by using information obtained during the computations to select a relatively small number of intervals to be subdivided.

Skelboe's method produces a sequence of values $(y_n)_{n=1}^{\infty}$ which converges under certain conditions to the *left endpoint* of $\tilde{f}(X)$ for an $X \in I^m$, on which f is defined. Analogously, the method may be applied to $-f$ to get the right endpoint of $\tilde{f}(X)$.

The following example of Skelboe (1974) describes the essential features of his method in the one-dimensional case, $m = 1$. Let $X \in I$, $f: X \to R$, and $F: I(X) \to I$ be an inclusion function for f which need not be inclusion isotone.

If the subdivision method of Section 4.1 is applied to produce refinements $S_n(X)$ for $n = 1, 2, 4, \ldots$ then a sequence of intervals $F(X_j^n) = [a_j^n, b_j^n]$, $j = 1, 2, \ldots, n$ are computed where X_j^n is the jth interval of that subdivision for which $w(X_j^n) = w(X)/n$ (see Fig. 4.1) and where we have suppressed the coordinate index used earlier in this chapter in the notation X_{ij} since $i = 1$.

In Skelboe's procedure an interval X_i^n is subdivided only if a_i^n is the lowest value in a certain ordered list of endpoints. This list has a key role in the procedure which is best seen by describing the first steps of the procedure.

1. (i) Subdivide $X = X_1^1$ into X_1^2, X_2^2.

   (ii) Compute the values a_1^2, a_2^2.

   (iii) Arrange these in a linearly ordered list (see Fig. 4.1), $L^2 = (a_1^2, a_2^2)$ such that $a_1^2 \leq a_2^2$.

   (iv) Select X_1^2 for further processing (since a_1^2 is the first member of the list L^2).

2. (i) Subdivide X_1^2 into X_1^4, X_2^4.
   (ii) Compute a_1^4, a_2^4.
   (iii) Delete a_1^2 from the list L^2 and insert a_1^4, a_2^4 on the list such that the linear order is kept (see Fig. 4.1) thus obtaining
   $L^4 = (a_2^2, a_1^4, a_2^4)$ with $a_2^4 \leq a_1^4 \leq a_2^2$.
   (iv) Select X_2^4 for further processing (since a_2 is the first member of the list L^4).

3. (i) Subdivide X_2^4 into X_3^8, X_4^8.
   (ii) Compute a_3^8, a_4^8.
   (iii) Delete a_2^4 from the list L^4 and insert a_3^8, a_4^8 on the list such that the linear order is kept thus obtaining ...

and so on.

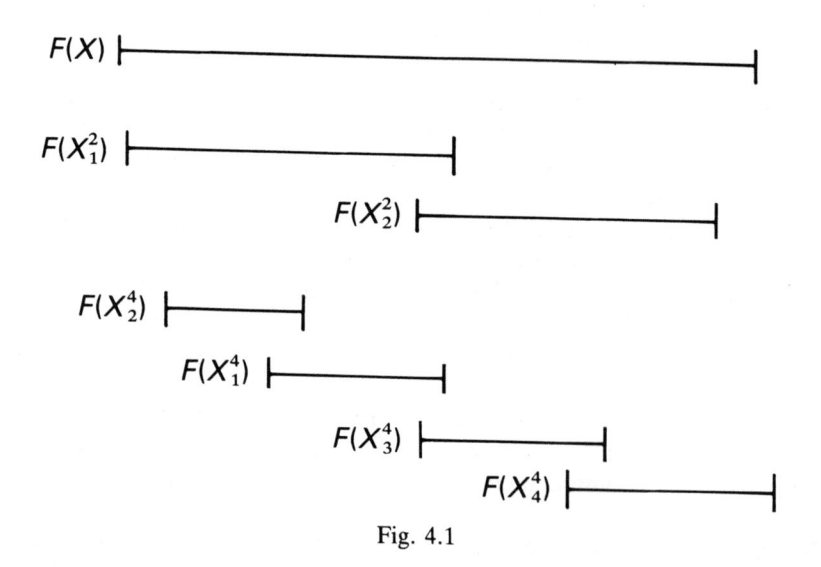

Fig. 4.1

The interval X_2^2 for example is only subdivided if a_2^2 is the first member of some list L^n (this will never happen in the situation shown in Fig. 4.1).

Continuing the procedure which is initiated by the steps 1, 2, and 3 we get an *infinite* sequence of values (y_n), $n = 2, 4, 8, 16, \ldots$ where y_n is the first element of the list L^n.

Theorem 4.4 Let $X \in I^m$, $f: X \to R$ be continuous and $F: I(X) \to I$ an inclusion function for f which converges to f, that is

$w[F(Y)] \to 0$ if $w(Y) \to 0$.

Then the sequence (y_n) of *left endpoints generated by Skelboe's method converges to the left endpoint of $\tilde{f}(X)$.*

Proof Let $y^* = \min \tilde{f}(X)$. Then, by construction,

$$y_n \leqslant y^*.$$

If Y_n is the interval which determines y_n, that is,

$$y_n = \min F(Y_n),$$

then $y_n, y^* \in F(Y_n)$. Since $w(Y_n) \to 0$ by construction, also $w[F(Y_n)] \to 0$ and therefore $y_n \to y^*$. $\square$

We now give some hints for the practical realization of Skelboe's procedure.

Remark 4.4 In order to obtain a termination criterion one may use the fact that $y^* \in F(Y_n)$ (see the proof of Theorem 4.4). If one calculates $F(Y_n)$ or $w[F(Y_n)]$ at each step (or after some initial steps) then the termination criterion could be

$$w[F(Y_n)] < \varepsilon.$$

This means that the absolute error is required to be smaller than ε. (There is no additional computational effort required to compute $w[F(Y_n)]$ or $\max F(Y_n)$ if the centred form is used.) An estimate of $w[F(Y_n)]$ may also be obtained via $w(Y_n)$ if a Lipschitz constant is known for F.

Remark 4.5 If F is inclusion isotone then the sequence (y_n) is monotonically increasing, that is

$$y_{n_1} \leqslant y_{n_2} \quad \text{whenever } n_1 \leqslant n_2.$$

In this case the equality

$$y_{n_1} = y_{n_2} \quad \text{for } n_1 \neq n_2 \tag{4.4}$$

may theoretically happen any number of times. If now the subdivision method is executed on a computer, then instead of y_n only numerical approximations $\tilde{y}_n$ are calculated. (It is clear that left rounding should be used in the numerical calculations.) The equality

$$\tilde{y}_{n_1} = \tilde{y}_{n_2} \quad \text{where } n_1 \neq n_2 \tag{4.5}$$

may again happen. The equality (4.5) may on the one hand be caused by property (4.4) and on the other, it may indicate that an improvement due to further subdivision is swamped by rounding errors such that it is pointless to continue the calculation. It is therefore not opportune to use this equality as a termination criterion as is done by several authors since it is not known by numerical calculation which of the two reasons is responsible for the equality (4.5). However, Asaithambi–Zuhe–Moore (1982) pose a class of functions where the equality (4.5) can only be caused by rounding errors.

Remark 4.6 If the list L^n becomes too long then it may be possible to remove the members of the tail from the list. Let y_n be the first member of the list, let Y_n be the related subinterval such that $F(Y_n) = [y_n, z_n]$ for some z_n. Then we may delete a member y from the list and also the related subinterval Y from the further processing if the condition

$$z_n < y$$

is satisfied since Y will never be used to approximate the lower bound. The same holds for all members of the list behind y.

The following version of Skelboe's procedure is due to Moore (1979). It subdivides the intervals X at each step only in one coordinate direction which changes cyclically. For instance, if $X = (X_1, ..., X_m)$ is subdivided in the ith direction, then we get $(X_1, ..., X'_i, ..., X_m)$ and $(X_1, ..., X''_i, ..., X_m)$ where $X_i = X'_i \cup X''_i$. The next subdivisions are along the $(i + 1)$st,..., mth, 1st,..., directions and so on.

The cyclic bisection algorithm (not including a termination criterion)
(1) Set $b_0 = \min F(X)$; set $Y = X$.
(2) Initialize *list* with the pair (Y, b_0).
(3) Set coordinate index $i = 1$.
(4) Bisect (subdivide) Y in coordinate direction i: $Y = Y_1 \cup Y_2$.
(5) Set $b_1 = \min F(Y_1)$ and $b_2 = \min F(Y_2)$.
(6) Remove (Y, b_0) from the list (that is, the first pair of the list).
(7) Enter the pairs (Y_1, b_1) and (Y_2, b_2) on the list in proper order (such that the second members of the pairs ascend).
(8) Denote the first pair of the list by (y, b_0).
(9) Cycle i (if $i < m$, then replace i by $i + 1$ else by 1).
(10) Go back to (4).

A new version of this algorithm which has some improvements for a practical computation can be found in Asaithambi–Zuhe–Moore (1982). Instead of the cycling bisection the bisecting is done at each step in the coordinate direction where the region has maximum width. This is done since these authors report simple two-dimensional examples where the cyclic choice of coordinate directions for bisection produces a sequence of thiner and thinner slit-shaped regions so that the method does not appear to converge. Further, this algorithm is enriched with some other improvements which will not be discussed here, see also Moore (1976). Finally, a termination criterion is provided that is valid for rational functions.

CHAPTER V

Optimality of the standard centred forms

The standard centred form of a rational function f over an interval X has turned out to be a good approximation for the range $\bar{f}(X)$ judging from many practical examples and tests. This chapter will demonstrate the *reason* for the good numerical results, that is, it will be shown that the standard centred form is an optimal approximation of $\bar{f}(X)$. This result is due to Ratschek–Rokne (1980a). The underlying concept of an optimal approximation is not the one that is commonly used, but it has constructive features which are extensively discussed in Ratschek (1980b). There are also some connections between this concept and the ideas of Traub–Woźniakowski (1980). For simplicity, we will only treat functions of one variable.

5.1 COMPUTABLE APPROXIMATIONS

Let **F** be a class of interval valued functions. A partial operator

$$\alpha: \mathbf{F} \times I \to I$$

is said to be an *including approximation for the ranges of the functions of* **F** (abbreviated: an *approximation for* **F**) if

$$\alpha(f, X) \supseteq \bar{f}(X) \quad \text{for all } f \in \mathbf{F}, X \in I \tag{5.1}$$

provided the terms $\alpha(f, X)$ and $\bar{f}(X)$ are defined. (The attribute 'partial' indicates that it is possible that $\alpha(f, X)$ is not always defined.)

In this chapter we will only admit approximations which carry constructive features. That is, if α is such an approximation for **F** then there shall exist some formula, algorithm, or instruction, etc., that evaluates the inclusion $\alpha(f, X)$ of $\bar{f}(X)$ for each $f \in \mathbf{F}$ and $X \in I$. Two conclusions may be drawn from this idea:

(i) There is some kind of algorithm $\hat{\alpha}$ which is associated with the approximation α and which has the purpose of evaluating $\alpha(f, X)$. The class of allowable operations or steps which may be allowed to be used is fixed and will depend on the actual case. For instance, this class can contain interval operations, max, min, comparisons, logical operations, definition by cases (if-statements), or also nth root, exp, log, sin, etc. Those who are interested in computer applications will certainly admit the usual computer instructions and statements. The present discussion only admits the class consisting of the four interval arithmetic operations.

(ii) For evaluating $\alpha(f, X)$, the algorithm $\hat{\alpha}$ manipulates only a finite set of information. This requirement is in accordance with the usual circumstances in computers and principles of formalized logic (Hilbert's finitist standpoint). The finite set of information enters the algorithm $\hat{\alpha}$ via the (finitely many) input parameters of $\hat{\alpha}$. The finite set of information will be presented by values for f: The *values* for (the functions of) **F** are understood to be finitely many mappings W_i: **F** $\to I$, $i = 1,...,k$. The mappings W_i can be constructive or not, they can be estimations, results of exact or numerical calculations, or information about any properties of the functions of **F**, etc.

An approximation α for **F** is said to be *dependent on the values* $W_1,...,W_k$ if there exists an interval function $\hat{\alpha}: D \to I$, where $D \subseteq I^{k+1}$ such that

$$\alpha(f, X) = \hat{\alpha}(W_1(f),..., W_k(f), X)$$

for all $f \in$ **F** and all $X \in I$, provided that $\alpha(f, X)$ is defined. If $\hat{\alpha}$ is a rational function, that is, only the four interval arithmetic operations are used, then α is called *computable from the values* $W_1,...,W_k$.

This concept of dependence of the calculation on some restricted information for a function f, and not on all the information about a function f, is very realistic and occurs practically, for example in physical observations, measurements, storage limitations of computers, etc.

The standard centred form of a rational function $f = p/q$ on X depends on the values $p(c),..., p^{(n)}(c)$, $q(c),..., q^{(n)}(c)$, and $H = X - c$, where $c = m(X)$. Therefore we demand that the approximations we admit for a comparison with the standard centred form shall depend on these data, since a reasonable comparison between various kinds of approximations is certainly only then possible if for all these approximations the same information, in our case the same data, is available. Clearly, if the information for one approximation consists of the data cited above, and the information for a second approximation consists of the lowest and the greatest values of f over X, then the second approximation will be the better approximation because with the two given values one knows the exact range immediately.

5.2 APPROXIMATIONS FOR THE RANGE

We are now going to apply the ideas developed in the previous section, to the standard centred form. We need:
(i) A class **F** of functions for which the range is to be evaluated; (ii) the values of the functions of **F** that are admitted or known for further calculation; (iii) the approximations that are admitted, that is, the formula that will provide inclusions for the range, using only the data of the functions of **F**.

Let **F** be the class of polynomial quotients p/q, where p and q are polynomials of degree at most n in one real variable. *Important*: We will identify two quotients p/q and p_1/q_1, iff $p = p_1$ and $q = q_1$. This requirement is necessary because on the one hand, $f = p/q$ and $f_1 = p_1/q_1$ have in general different data and different centred forms, even if f and f_1 are identical as functions, and on the other hand, because in practical calculations it is not always decidable if $f = f_1$ or equivalently, if p and q have a common polynomial factor.

Now, the values for the functions of **F** are the assignments

$$W_k: \mathbf{F} \to R \quad \text{for } k = 1, \ldots, 2n + 2$$

where

$$W_1(f) = p(c), \ldots, W_{n+1}(f) = p^{(n)}(c),$$
$$W_{n+2}(f) = q(c), \ldots, W_{2n+2}(f) = q^{(n)}(c)$$

for $f = p/q \in \mathbf{F}$ and $c = m(X)$. Therefore the values depend on X, that is, the actual argument of the range function. For the sake of compactness we let

$$W = (W_1, \ldots, W_{2n+2}),$$
$$Wf = (W_1(f), \ldots, W_{2n+2}(f)).$$

An arithmetic formula which may be used to evaluate the standard centred form (of first order) for the functions of **F** in the sense of Section 5.1 is given as

$$s(u, H) = \frac{u_1}{u_{n+2}} + \frac{\sum_{k=1}^{n}(u_{k+1} - u_1 u_{n+k+2}/u_{n+2})H^k/k!}{\sum_{k=0}^{n} u_{n+k+2} H^k/k!}$$

where the powers H^k are evaluated in the simple form and where $u = (u_1, \ldots, u_{2n+2})$ is a variable over R^{2n+2}. That means, given a rational function $f \in \mathbf{F}$ and an interval X such that the standard centred form of f on

X, $F(X)$, is defined, then

$$s(Wf, H) = F(X)$$

where $H = X - c$ and $c = m(X)$.

Thus s can be seen as some kind of an algorithm with $u_1, \ldots, u_{2n+2}$, and H as input variables, that is, s is an underlying rational formula for the standard centred form and the operator

$$(f, X) \to F(X)$$

is a computable approximation from the values W defined previously. Henceforth we will not distinguish between an approximation for **F** and the underlying formula in the sense of Section 5.1, because there is no danger of confusion.

The character of the standard centred form s suggests that it should be compared to approximations for **F** of a similar character.

Let the function $\beta\colon R^{2n+2} \times I \to I$ be representable in the form

$$\beta(u, H) = \left(\sum_{k=0}^{m} c_k(u)H^k\right) \bigg/ \left(\sum_{k=0}^{m} d_k(u)H^k\right) \tag{5.2}$$

where H^k is evaluated in the simple form, m is a non-negative integer, and c_k, d_k are real-valued rational functions over R^{2n+2}. Let **B** be the class of all such functions β which satisfy for each $f \in \mathbf{F}$ and $X \in I$ the inclusion condition

$$\beta(Wf, H) \supseteq \tilde{f}(X) \tag{5.3}$$

where $H = X - c$ and $c = m(X)$ provided the terms are defined. We notice that **B** is a class of approximations for **F** which are computable from the values W.

We are now going to compare the standard centred form with the approximations from **B** and it will be shown that no approximation from **B** is 'better' than the standard centred form, that is, the approximation s is an *optimal* approximation in the class $\mathbf{B} \cup \{s\}$. The concept of 'better' is made precise in the following manner:

If β and γ are approximations of $\mathbf{B} \cup \{s\}$, then β is said to be *better* than γ if

$$\beta(Wf, H) \subseteq \gamma(Wf, H) \tag{5.4}$$

for all $f \in \mathbf{F}$, $X \in I$, where $H = X - c$ and $c = m(X)$, provided both sides of the inclusion of (5.4) are defined.

If β is an approximation of **B** and if we set $H = 0$, then we get from (5.2) and (5.3)

$$c_0(Wf)/d_0(Wf) = \tilde{f}([c, c]) = f(c) = W_1(f)/W_{n+2}(f).$$

Sec. 5.2]   **Approximations for the Range**   107

Since the previous equality holds for all $f \in \mathbf{F}$, provided $\beta(Wf, 0)$ is defined, it follows that $c_0(u)/d_0(u) = u_1/u_{n+2}$. Since we can multiply the numerator and the denominator of β with reals without changing the functional assignment, we normalize β such that $d_0(u) = u_{n+2}$. Thus, we get for each $\beta \in \mathbf{B}$,

$$\left. \begin{array}{ll} c_0(u) = u_1, & d_0(u) = u_{n+2}, \\ c_0(Wf) = p(c) & d_0(Wf) = q(c) \quad \text{for } f = p/q \in \mathbf{F} \end{array} \right\} \quad (5.5)$$

5.3 A SPECIAL CASE

In this section the approximations β of $\mathbf{B}$ will be applied to a certain subclass $\mathbf{G} \subset \mathbf{F}$ and then, in Section 5.4, we will obtain the final results.

Let $\mathbf{G}$ be the calss of all functions $p/q \in \mathbf{F}$ with

$$p^{(k)}(c) \in 2^{2n-2k}[1, 2] \quad \text{for } k = 0,\ldots,n \tag{5.6}$$

and where

$$\left. \begin{array}{l} 0 < q(c) < 1, \\ -1 \leq q^{(k)}(c) \leq 0, \quad \text{for } k = 1,\ldots,n-1, \\ q^{(n)}(c) \leq -n! \end{array} \right\} \tag{5.7}$$

These conditions permit us to make some conclusions about the monotonicity of $f = p/q$ and the location of the zeros of p and q, as can be seen in the proofs of Lemmas 5.1 and 5.2.

Lemma 5.1 *If $p/q \in \mathbf{G}$ then the polynomials $\bar{p}(z) = \sum_{k=0}^{n} p^{(k)}(c) z^k/k!$ and $\bar{q}(z) = \sum_{k=0}^{n} q^{(k)}(c) z^k/k!$ have no common polynomial divisors.*

Proof We will show that $\bar{p}$ and $\bar{q}$ have no common (real or complex) zeros. It follows from (5.6) that the coefficients of $\bar{p}$ are positive and that $[p^{(k)}(c)/k!]/[p^{(k+1)}(c)/(k+1)!] \geq 2$. By a theorem about polynomials we get $|\xi| \geq 2$ for each zero ξ of $\bar{p}$, cf. Marden (1966, p. 137). It follows from (5.7) that

$$\frac{q^{(k)}(c)/k!}{q^{(n)}(c)/n!} \leq \frac{1}{k!} \leq 1, \text{ for } k = 0,\ldots, n-1.$$

By a theorem about polynomials we get $|\xi| < 2$ for each zero ξ of $\bar{q}$, cf. Marden (1966, p. 123). □

Let again $F(X)$, $P(X)$, and $Q(X)$ denote the standard centred forms of f, p and q over $X = c + H$, $H = [-z, z]$, where $c = m(X)$, $z = w(X)/2$, and p and q are polynomials of degree n such that (5.6) and (5.7) hold when we write $f = p/q \in \mathbf{G}$.

Lemma 5.2 Let β be an approximation for **F** of **B**. If $\beta(Wf, h) \subseteq F(X)$ for all $f = p/q \in \mathbf{G}$ and all $X \in I$, provided the occurring intervals are defined, then

$$\beta(Wf, H) = P(X)/Q(X)$$

and

$$\left. \begin{array}{l} c_k(u) = u_{k+1}/k! \\ d_k(u) = u_{n+k+2}/k! \end{array} \right\} \text{ for } u \in R^{2n+2} \text{ and } k = 0, \ldots, n. \qquad (5.8)$$

Proof We write f in the form

$$f(x) = f(c) + \frac{t_1(x-c) + \ldots + t_n(x-c)^n/n!}{q(c) + \ldots + q^{(n)}(c)(x-c)^n/n!}$$

see (2.1) and (2.6). The numbers $t_k = p^{(k)}(c) - f(c)q^{(k)}(c)$ are non-negative for $k = 1, \ldots, n$. Thus, f is monotonically increasing for $x \geq c$ and we get

$$\tilde{f}([c, c+z]) = f(c) + \left[0, \frac{t_1 z + \ldots + t_n z^n/n!}{q(c) + \ldots + q^{(n)}(c)z^n/n!} \right].$$

Comparing this interval with $F(X)$ we see that for all $z \geq 0$ the right endpoints of these both intervals are equal. Since

$$\tilde{f}([c, c+z]) \subseteq \tilde{f}(X) \subseteq \beta(Wf, H) \subseteq F(X) \quad \text{ for all possible } X$$

the right endpoints of $\beta(Wf, H)$ coincide with the right endpoints of $F(X)$, that is, using (5.5) and writing c_k and d_k instead of $c_k(Wf)$ and $d_k(Wf)$ we get:

$$\frac{p(c) + \sum_{k=1}^{m} |c_k| z^k}{q(c) - \sum_{k=1}^{m} |d_k| z^k} = f(c) + \frac{\sum_{k=1}^{n} t_k z^k/k!}{\sum_{k=0}^{n} q^{(k)}(c) z^k/k!}$$

$$= \frac{\sum_{k=0}^{n} p^{(k)}(c) z^k/k!}{\sum_{k=0}^{n} q^{(k)}(c) z^k/k!}$$

The second of the two equations arises by arranging the partial fraction of the left side of the second equation as one fraction. By Lemma 5.1, the numerator and the denominator of the right quotient have no common divisors. Furthermore, the absolute coefficients of the denominators of the quotients are equal and the equation above holds for some proper interval $[0, z_0]$, thus the following comparison of coefficients is possible:

$$\begin{array}{ll} |c_k| = p^{(k)}(c)/k! & \text{for } k = 1, \ldots, n, \\ -|d_k| = q^{(k)}(c)/k! & \text{for } k = 1, \ldots, n, \\ c_k = d_k = 0 & \text{for } k > n. \end{array}$$

Inserting these values for the coefficients we get $\beta(Wf, H) = P(X)/Q(X)$. Since the coefficients c_k and d_k for $k \neq 0$ only occur in connection with the symmetrical interval H, we can assume $c_k, d_k \geq 0$ for $k \neq 0$.

It remains to verify (5.8). For this it is sufficient to consider c_k for some fixed k. First, we consider the proper parallelepiped $K = \prod_{i=1}^{2n+2} A_i$, where A_i is given by

$$A_{1+k} = 2^{2n-2k}[1, 2] \quad \text{for } k = 0,\ldots,n,$$
$$A_{n+1+k} = [-1, 0] \quad \text{for } k = 1,\ldots,n-1,$$
$$A_{n+2} = [0, 1],$$
$$A_{2n+2} = [-2, -1]\, n!,$$

see (5.6) and (5.7). Then the equation $c_k(u) = u_{k+1}/k!$ holds in K, because for every $u \in K$ there exists an $f = p/q \in \mathbf{G}$ with $u_1 = p(c),\ldots, u_{2n+2} = q^{(n)}(c)$, and $c_k(u) = c_k(Wf) = p^{(k)}(c)/k! = u_{k+1}/k!$. Since $c_k(u) = u_{k+1}/k!$ holds for all $u \in K$ and $c_k(u)$ is a rational function in u, the equality $c_k(u) = u_{k+1}/k!$ holds for all $u \in R^{2n+2}$. □

5.4 THE GENERAL CASE

The lemmas of Section 5.3 will be applied to get a contradiction by assuming that an approximation exists that is better than the centred form:

Theorem 5.1 *Let β be an approximation for $\mathbf{F}$ of the class $\mathbf{B}$. Then β is not better than the standard form.*

Proof (by contradiction). Let us assume that for all $f \in \mathbf{F}$ the inclusion

$$\beta(Wf, H) \subseteq F(X)$$

holds. Then for all $f \in \mathbf{G}$ using Lemma 5.2 we get $c_k(u) = u_{k+1}/k!$ and $d_k(u) = u_{n+k+2}/k!$ for all u and $k = 0,\ldots, n$. It follows that

$$c_k(Wf) = p^{(k)}(c)/k! \quad \text{and} \quad d_k(Wf) = q^{(k)}(c)/k!$$

and finally $\beta(Wf, H) = P(X)/Q(X)$ for every $f = p/q \in \mathbf{F}$.

There exists an $f = p/q \in \mathbf{F}$, however, such that $F(X) \subset \beta(Wf, H)$ is a proper inclusion, and the contradiction is achieved. It suffices to choose $f(x) = (8 + x)/(4 + x)$, $c = 0$ and $H = X = [-1, 1]$. Then we get

$$p(0) = 8,\ p'(0) = 1,\ q(0) = 4,\ q'(0) = 1,\ t_1 = -1$$

and

$$\frac{P(X)}{Q(X)} = \left[\frac{21}{15}, \frac{45}{15}\right] \supset F(X) = \left[\frac{25}{15}, \frac{35}{15}\right]. \quad \Box$$

Remark 5.1 The standard centred form is not representable as a quotient

$$(c_0 + \ldots + c_n H^n)/(d_0 + \ldots + d_n H^n).$$

(Assuming such a representation and denoting it by β, the proof of Theorem 5.1 shows how to get a contradiction.)

Remark 5.2 Since the standard centred form is an expression of the form

$$c_0 + (c_1 H + \ldots + C_n H^n)/(d_0 + \ldots + d_n H^n)$$

it is reasonable to ask if approximations of this form can be better than the centred form. It can also be shown that it is not possible. A proof of this assertion is similar to the proofs in this chapter and is therefore omitted here.

Remark 5.3 If an approximation is optimal for the class **F**, then it need not be the case for a subclass $\mathbf{F}_0 \subset \mathbf{F}$, cf. Ratschek (1980b). Therefore, a special discussion for the class of polynomials of **F**, which is a very important subclass of **F**, is reasonable. Such a discussion appears in Ratschek–Rokne (1981). There it is shown that the standard centred form is also optimal for this subclass and additionally, the optimality is saved if the powers H^k are evaluated in the extended form. The optimality of the standard centred form for polynomials with respect to the interval width is also proven there. This means that the 'better' relation (5.4) is replaced by the relation

$$w[\beta(Wf, H)] \leq w[\gamma(Wf, H)].$$

Remark 5.4 We keep in mind that the standard centred forms of higher order manipulate the same data as the standard centred form and are better approximations for **F** than the standard centred form. The forms of higher order have, however, a greater computational complexity than the form of first order. Thus the class **B** of approximations can be seen as a boundary with respect to the complexity and it is not possible to represent the standard centred forms of higher order by an approximation of **B**.

CHAPTER VI

Other inclusions for the range of a function

In this chapter we discuss a number of methods for including the range of a function that are either not based on the idea of a centred form or that use the centred form in a manner different from the previous use. These methods have in some cases only restricted applicability. Advantages and disadvantages of these methods are also treated. We also discuss an application of the centred form to iteration operators. Finally the remainder and interpolation forms are considered. These give convergence of higher than second order.

6.1 QUOTIENTS OF CENTRED FORMS

Let $f = f_1/f_2$ be the quotient of two real functions f_1 and f_2 defined on an interval X and let $F_1(Y)$ and $F_2(Y)$ be centred form functions for f_1 and f_2 respectively for $Y \in I(X)$. The quotient

$F_1(Y)/F_2(Y)$ for $Y \in I(X)$

is an inclusion function for the range function $\bar{f}(Y)$ although it is not necessarily a centred form function. Such quotients are investigated in this section. In the case of rational functions it turns out that the evaluation of $F_1(Y)/F_2(Y)$ requires fewer operations than the evaluation of the standard centred form. The disadvantage is that the quadratic convergence of $F_1(Y)/F_2(Y)$ to $\bar{f}(Y)$ may only be shown to be valid for a very restricted class of functions.

Quotients of centred forms were first discussed by Alefeld–Rokne (1981) for the standard centred form as well as for the mean value form. As a particular example consider $f: X \to R$ defined by

$f(x) = 1/q(x)$

where $q(x)$ is a polynomial. Clearly, in order to compute the standard centred form for f over $X \in I$ more operations are required than for the form

$$1/Q(X)$$

where $Q(X)$ is the standard centred form for q over X. This particular case turns out to be always quadratically convergent.

The result of Alefeld–Rokne (1981) in one variable was as follows: let p and q be polynomials and let $f = p/q$ be defined on $X \in I$. Let

$$F(Y) = P(Y)/Q(Y) \quad \text{for all } Y \in I(X) \tag{6.1}$$

where $P(Y)$ and $Q(Y)$ are the standard centred forms for p and q on Y provided $0 \notin Q(Y)$. If it is assumed that

$$f(c)\,p'(c)\,q'(c) \leq 0 \quad \text{for all } c \in X, \tag{6.2}$$

then it can be shown that F is quadratically convergent to $\tilde{f}$, that is, there exists an α such that

$$|F(Y), \tilde{f}(Y)| \leq \alpha w(Y)^2 \quad \text{for all } Y \in I(X).$$

Clearly, the condition (6.2) is quite restrictive and hence only satisfied by few functions. Furthermore, it is numerically quite difficult to check whether (6.2) is satisfied or not. The quotient form is therefore not very suitable for use with a subdivision method unless (6.2) is known to be satisfied *a priori*. If, however, it is only required to produce a rough estimate of $\tilde{f}(Y)$ for a single Y then the quotient form (6.1) is a candidate because of its lower computational complexity.

In this section we consider the quotient of *general* centred forms. The results of Alefeld–Rokne (1981) will then follow as special cases of the general results.

Let $X \in I^m$ and let the real functions f_1, f_2, and $f = f_1/f_2$ be defined on X. For $i = 1, 2$, let $F_i(Y) = f_i(c) + S_i(Y)$ be a centred form function for f_i with $S_i(Y) \subseteq \sum_{\rho=1}^{r} (Y - c) \cdot G_i^\rho(Y)$ and with a common point evaluation function $c = \alpha(Y) \in Y$. Assuming $0 \notin F_2(Y)$ we set $F(Y) = F_1(Y)/F_2(Y)$ for all $Y \in I(X)$. We then prove the following results:

Theorem 6.1

(i) $\tilde{f}(Y) \subseteq F(Y)$ for all $Y \in I(X)$,

(ii) if G_1^ρ, G_2^ρ, and $1/F_2$ are bounded, then $F(Y)$ converges linearly to $\tilde{f}(Y)$ if $w(Y)$ tends to 0.

Proof

(i) Clear, since $f(x) = f_1(x)/f_2(x) \in F_1(Y)/F_2(Y)$ if $x \in Y$.

(ii) In the following estimation we use the formulas (1.6):

Sec. 6.1]   Quotients of Centred Forms   113

$$w[F(Y)] \leq |F_1(Y)| w[1/F_2(Y)] + w[F_1(Y)] |1/F_2(Y)|$$
$$\leq |F_1(Y)| |1/F_2(Y)|^2 w[F_2(Y)] + |1/F_2(Y)| w[F_1(Y)]$$

Since the functions G_1^ρ and G_2^ρ are bounded, we get

$$w[F_i(Y)] \leq Kw(Y) \text{ for } i = 1, 2 \text{ and all } Y \in I(X)$$

for some constant K, see the proof of Theorem 3.1.

The remaining terms are bounded by the assumption such that there exists a constant L with

$$w[F(Y)] - w[\tilde{f}(Y)] \leq w(F(Y)) \leq Lw(Y) \qquad \text{for all } Y \in I(X). \qquad \square$$

The condition that $1/F_2$ is bounded in Theorem 6.1 can easily be satisfied. For instance, it holds if F_2 is continuous since $I(X)$ is compact by Lemma 1.8.

Theorem 6.2 Let the functions f_1 and f_2 be twice differentiable, let the functions $G_i = (G_{i1}, \ldots, G_{im})$ be Lipschitz on $I(X)$, and let $c = m(Y)$ be the midpoint function. If

$$f_i(c) \in \sum_{\rho=1}^{r} G_{iv}^\rho(Y), \frac{\partial}{\partial x_v} v = 1, \ldots, m, \qquad i = 1, 2, \qquad (6.3)$$

and

$$f(x) \frac{\partial f_1(x)}{\partial x_v} \frac{\partial f_2(x)}{\partial x_v} \leq 0 \qquad \text{for } v = 1, \ldots, m \qquad (6.4)$$

then $F(Y)$ converges quadratically to $\tilde{f}(Y)$.

Proof Since the functions f_i are twice differentiable they can be written in the form

$$f_i(x) = a_i + b_i \cdot (x - c) + O(\|x - c\|^2)$$

for $x \in Y \in I(X)$ using Taylor's formula. The coefficients of this representation depend on Y via the developing function $c = m(Y)$, that is,

$$a_i = a_i(Y) \in R, \quad b_i = b_i(Y) \in R^m \qquad (i = 1, 2).$$

The assumptions (6.3) and (6.4) are then written as

$$b_{iv} \in \sum_{\rho=1}^{\mu} G_{iv}^\rho(Y) \qquad v = 1, \ldots, m, \qquad \text{and} \qquad i = 1, 2 \qquad (6.5)$$

$$(a_1/a_2) b_{1v} b_{2v} \leq 0 \qquad \text{for } v = 1, \ldots, m. \qquad (6.6)$$

Using the boundedness of the functions G_i^ρ and the easily verifiable formula

$$1/A = A/a^2 + t^2/(a^2 A)$$

which holds for intervals $A = a + [-t, t]$ with $0 \leq t < |a|$, we get the following inclusion for $F(Y)$,

$$F(Y) = F_1(Y)/F_2(Y) \subseteq (a_1 + T_1)/(a_2 + T_2)$$
$$= (a_1 + T_1)(a_2 + T_2 + t_2^2/(a_2 + T_2))/a_2^2$$
$$\subseteq a_1/a_2 + (a_1/a_2^2) T_2 + T_1/a_2 + O(w(Y)^2)$$

where the definition

$$T_i = [-t_i, t_i] = \sum_{\rho=1}^{r} (Y - c) \cdot G_i^\rho(Y), \; i = 1, 2 \text{ is used.}$$

is used. Using (2.13) we then estimate the following upper bound for the width of $F(Y)$:

$$w[F(Y)] \leq \frac{|a_1|}{a_2^2} \sum_{\rho=1}^{r} \sum_{v=1}^{m} |G_{2v}^\rho(Y)| \, w(Y_2)$$

$$+ \frac{1}{|a_2|} \sum_{\rho=1}^{r} \sum_{v=1}^{m} |G_{1v}^\rho(Y)| \, w(Y_v) + O(w(Y)^2)$$

$$= \sum_{v=1}^{m} w(Y_v) \left[\frac{|a_1|}{a_2^2} \sum_{\rho=1}^{r} |G_{2v}^\rho(Y)| + \sum_{\rho=1}^{r} |G_{1v}^\rho(Y)| \right]$$

$$+ O(w(Y)^2). \tag{6.7}$$

In order to get an upper bound for $w[F(Y)] - w[\tilde{f}(Y)]$ we now construct a lower bound for $w[\tilde{f}(Y)]$. Using the series development of $1/(1 + z)$ for $|z| < 1$ we get for $x \in Y$

$$f(x) = [a_1 + b_1 \cdot (x - c) + O(\|x - c\|^2)]/[a_2 + b_2 \cdot (x - c) + O(\|x - c\|^2)]$$
$$= \alpha_0 + \alpha_1 \cdot (x - c) - \beta_1 \cdot (x - c) + O(\|x - c\|^2) \tag{6.8}$$

where $\alpha_0 = a_1/a_2 \in R$, $\alpha_1 = b_1/a_2 \in R^m$, and $\beta_1 = (\alpha_0/a_2) b_2 \in R^m$. If $y, z \in Y$, then

$$f(y) - f(z) = \alpha_1 \cdot (y - z) - \beta_1 \cdot (y - z) + O(w(Y)^2)$$

$$= \sum_{v=1}^{m} [\alpha_{1v}(y_v - z_v) - \beta_{1v}(y_v - z_v)] + O(w(Y)^2).$$

Now we chose y and z such that

(i) $\alpha_{1v}(y_v - z_v) \geq 0$ for $v = 1,\ldots,m$.
(ii) $\beta_{1v}(y_v - z_v) \leq 0$ for $v = 1,\ldots,m$,
(iii) $|y_v - z_v| = w(Y_v)$ for $v = 1,\ldots,m$.

The conditions (i) and (ii) are reasonable since $\alpha_{1v}\beta_{1v} = b_{1v}b_{2v}a_1/a_2^3 \leq 0$ for $v = 1,\ldots,m$ by (6.6). Condition (iii) means that $Y_v = y_v \vee z_v$ and further that $Y = y \vee z$. Then, $f(y) - f(z)$ can be written as follows,

$$f(y) - f(z) = \sum_{v=1}^{m} (|\alpha_{1v}| + |\beta_{1v}|)w(Y_v) + O(w(Y)^2). \quad (6.9)$$

If $f(y) - f(z) \geq 0$, then we can replace this difference by its absolute value in (6.9). If $f(y) - f(z) < 0$, then (6.9) implies that both the difference as well as the sum are $O(w(Y)^2)$. It follows in this case that also $|f(y) - f(z)| = O(w(Y)^2)$ and we get

$$w[\tilde{f}(Y)] \geq |f(y) - f(z)| = \sum_{v=1}^{m} (|\alpha_{1v}| + |\beta_{1v}|)w(Y_v) + O(w(Y)^2) \quad (6.10)$$

We finally apply the formula

$$||A| - |a|| \leq w(A) \quad \text{if } a \in A \in I$$

to the relations (6.5) and get the following estimations

$$\left| \frac{1}{|a_2|} \left| \sum_{\rho=1}^{r} G_{1v}^{\rho}(Y) \right| - |\alpha_{1v}| \right| \leq \frac{1}{|a_2|} \sum_{\rho=1}^{r} w\left[G_{1v}^{\rho}(Y) \right] \leq Kw(Y),$$

$$\left| \frac{|a_1|}{a_2^2} \left| \sum_{\rho=1}^{r} G_{2v}^{\rho}(Y) \right| - |\beta_{1v}| \right| \leq \frac{|a_1|}{a_2^2} \sum_{\rho=1}^{r} w\left[G_{2v}^{\rho}(Y) \right] \leq Lw(Y).$$

The constants K and L are independent of Y and arise since the G_{iv}^{ρ} are Lipschitz and since $1/a_2 = 1/f_2(c)$ is bounded for $c \in X$ due to the existence of $1/F_2$. The above estimations together with (6.7) and (6.10) leads to the desired quadratic convergence property, i.e.

$$w[F(Y)] - w[\tilde{f}(Y)] \leq \sum_{v=1}^{m} w(Y_v) [Kw(Y) + Lw(Y)] + O(w(Y)^2)$$

$$\leq m(K + L)w(Y)^2 + O(w(Y)^2). \quad \square$$

We now illustrate our results by some simple examples.

Example 6.1 Let $X = [-1, 1]$, $f(x) = (2 - x)/(2 + x)$, $F_1(Y) = 2 - c - (Y - c)$, and $F_2(Y) = 2 + c + (Y - c)$, where the point function $c = \alpha(Y) \in Y$ is arbitrary. One can easily check that the assumptions of part (iii) of Theorem 6.2 hold, such that quadratic convergence follows. In fact, if

$$Y = [u, v], \text{ then } \tilde{f}(Y) = \left[\frac{2 - v}{2 - u}, \frac{2 - u}{2 - v} \right] \text{ and } F(Y) = \tilde{f}(Y),$$

and quadratic convergence is trivially fulfilled.

Although condition (6.4) seems at first sight to be rather artificial and only necessary for the given proof of Theorem 6.2, the following example shows that without (6.4) the statement (iii) of the theorem is not true in general:

Example 6.2 Let $X = [-1, 1]$, $f(x) = (1 + x)/(2 + x)$, $F_1(Y) = 1 + c + (Y - c)$, $F_2(Y) = 2 + c + (Y - c)$, and $c = \alpha(Y) \in Y$ be arbitrary. One can easily check that condition (6.3) does not hold. Quadratic convergence is not given: if $Y = [u, v]$, then

$$\tilde{f}(Y) = \left[\frac{1+u}{2+u}, \frac{1+v}{2+v}\right]$$

and

$$F(Y) = \left[\frac{1+u}{2+v}, \frac{1+v}{2+u}\right]$$

In particular, we set $u = 0$ such that $v = w(Y)$ and get

$$w[F(Y)] - w[\tilde{f}(Y)] = v/2 = w(Y)/2.$$

Thus we have at most linear convergence in a special case, such that quadratic convergence cannot be expected in general.

Remark 6.1 In the discussion on the use of the mean-value theorem in Section 3.5 it was pointed out that it was often difficult to obtain a satisfactory inclusion for the derivative. This is not the case in Theorem 6.2 since it is simple to find an inclusion for the derivative of a function which satisfies (6.5).

6.2 THE CIRCULAR COMPLEX CENTRED FORM

In this section we consider the problem of bounding the range of a complex polynomial over a circular interval (= complex disc). This problem was discussed in Rokne–Wu (1982, 1983) where it was shown that the complex centred form is better than some other forms and that the radius [centre] of the complex centred form converges quadratically [linearly] to the radius [centre] of the smallest disc containing the domain as the diameter of the domain tends to zero. We follow that development here and we also prove the quadratic convergence of the just mentioned centres. Finally, Krawczyk's circular centred form for complex rational functions is described. Further circular centred forms are compared by Petković (1983).

The set of complex numbers is denoted by C. Let $c \in C$ and $\rho \in R$ such that $\rho \geq 0$. The set

$$Z = \{z: |a - c| \leq \rho\}$$

is called a *circular interval*. The set of circular intervals is denoted by $I(C)$. A circular interval $Z \in I(C)$ may be written as

$$Z = \langle c, \rho \rangle$$

and the notations $c = \text{mid } Z$ and $\rho = \text{rad } Z$ are used with their obvious interpretations. Let $\langle c, \rho \rangle \in I(C)$, let $Z_i = \langle c_i, \rho_i \rangle \in I(C)$; $i = 1, 2, \ldots, k$ and let $a \in C$. The following operations were defined by Gargantini–Henrici (1972):

$$a + Z = \langle a + c, \rho \rangle = \{a + z : z \in Z\}$$
$$aZ = \langle ac, |a|\rho \rangle = \{az : z \in Z\} \tag{6.11}$$
$$\sum_{i=1}^{k} Z_i = \langle \sum_{i=1}^{k} c_i, \sum_{i=1}^{k} \rho_i \rangle = \{\sum_{i=1}^{k} z_i : z_i \in Z_i, i = 1, \ldots, k\}$$
$$Z_1 Z_2 = \langle c_1 c_2, |c_1|\rho_2 + |c_2|\rho_1 + \rho_1 \rho_2 \rangle.$$

Unfortunately, the set $\{z_1 z_2 : z_i \in Z_i, i = 1, 2\}$ is in general not a disc. It is, however, a subset of $Z_1 Z_2$ which implies that the given operations remain within the aims of interval arithmetic as discussed in Sections 1.1 and 1.2. The *subdistributive law*

$$Z_1(Z_2 + Z_3) \subseteq Z_1 Z_2 + Z_1 Z_3 \tag{6.12}$$

is also valid and the equation

$$Z^n = \langle c^n, (|c| + \rho)^n - |c|^n \rangle \tag{6.13}$$

is easily proven using complete induction.

The complex circular arithmetic used in the sequel is the one defined by equations (6.11) having properties (6.12) and (6.13).

Let $a_i \in C$, $i = 0, 1, 2, \ldots, n$. Then $p(z) = \sum_{i=0}^{n} a_i z^i$, $z \in Z$, is a complex polynomial. We seek outer circular interval estimates for the set

$$\bar{p}(Z) = \{p(z) : z \in Z\}$$

for a $Z \in I(C)$. One such estimate is the circular interval

$$p_S(Z) = a_0 + a_1 Z + \ldots + a_n Z^n$$

which we will call the *power sum evaluation* of $p(z)$ over Z. It is nothing but the natural extension of $p(z)$ to Z. A second estimate is

$$p_H(Z) = (\ldots((a_n Z + a_{n-1})Z + a_{n-2})Z + \ldots + a_0) \tag{6.14}$$

which we will call the *Horner scheme evaluation* of $p(z)$ over Z. Clearly, $p_H(Z)$ is the natural interval extension of the Horner scheme representation of $p(z)$ to Z. For these two evaluations we have the following result.

Theorem 6.3 *Let $p(z) = \sum_{i=0}^{n} a_i z^i$ be a complex polynomial and let $Z = \langle c, \rho \rangle \in I(C)$. It then follows that $p_S(Z) = \langle c_S, \rho_S \rangle$ and $p_H(Z) = \langle c_H, \rho_H \rangle$ are given by*

$$c_S = p(c)$$
$$\rho_S = \sum_{i=1}^{n} |a_i|\,((|c| + \rho)^i - |c|^i), \qquad (6.25)$$

$$c_H = p(c),$$
$$\rho_H = \rho \sum_{i=1}^{n-1} \{(|c| + \rho)^{i-1} | \sum_{j=i}^{n-1} a_j c^{j-i} |\}. \qquad (6.16)$$

Furthermore, the inclusions

$$\bar{p}(Z) \subseteq p_H(Z) \subseteq p_S(Z) \qquad (6.17)$$

are valid.

Proof Using (6.11) and (6.13) it follows immediately that (6.15) holds. The proof of (6.16) is somewhat longer and proved using complete induction on the degree n of the polynomial. For $n = 1$ the assertion is obvious. Assume that (6.16) is valid for polynomials with degree less than n. Then if the degree of $p(z)$ is equal to n we obtain

$$p_H(Z) = p_H^*(Z) \cdot Z + a_0$$

where $p_H^*(Z)$ denotes the Horner scheme evaluation of the polynomial $p^*(z) = \sum_{i=1}^{n} a_i z^{i-1}$ over Z. Using the induction hypothesis for $p_H^*(Z)$ we obtain

$$p_H(Z) = p_H^*(Z)\, Z + a_0$$
$$= \langle p^*(c),\, \rho \sum_{i=1}^{n-1} \{(|c| + \rho)^{k-1} | \sum_{i=1}^{n-1} a_{j+1}\, c^{j-i}|\} \rangle \langle c, \rho \rangle + a_0$$
$$= \langle a_0 + c p^*(c),\, (|c| + \rho) \{\rho \sum_{i=1}^{n-1} [(|c| + \rho)^{i-1}$$
$$| \sum_{j=1}^{n-1} a_{j+1}\, c^{j-i} |]\} + \rho |p^*(c)| \rangle$$
$$= \langle a_0 + c \sum_{i=1}^{n} a_i\, c^{i-1},\, \rho \{ \sum_{i=2}^{n} [(|c| + \rho)^{i-1}$$
$$| \sum_{j=i-1}^{n-1} a_{j+1}\, c^{j-(i-1)} |] + | \sum_{j=i}^{n} a_j\, c^{i-1}|\} \rangle$$
$$= \langle p(c),\, \rho \sum_{i=1}^{n} \{(|c| + \rho)^{i-1} | \sum_{j=1}^{i} a_j\, c^{j-i}|\} \rangle.$$

This proves (6.16).

The inclusion $p_H(Z) \subseteq p_S(Z)$ follows from the subdistributive law (6.12) and the inclusion $\bar{p}(Z) \subseteq p_H(Z)$ is valid because of (6.11). □

Sec. 6.2] The Circular Complex Centred Form

A further possibility for estimating the range of a complex polynomial is given by the generalization of the standard centred form for real polynomials to the complex case. The *standard circular centred form function* (shortly, *circular centred form function* or *centred form function*) for a complex polynomial $p(z) = \sum_{i=0}^{n} a_i z^i$ over a circular interval Z_0 is a natural interval extension of the Taylor expansion

$$\sum_{i=0}^{n} p^{(i)}(c)(z-c)^i/i! \qquad (6.18)$$

of $p(z)$ at $c \in Z_0$, analogously to the real case. Let $I(Z_0)$ denote the set of all circular intervals $Z \subseteq Z_0$ and $c = c(Z) = \text{mid } Z$ the developing point function (we will always write c instead of $c(Z)$). Then the centred form function of p over Z_0, $p_C: I(Z_0) \to I(C)$ is defined by

$$p_C(Z) = \sum_{i=0}^{n} p^{(i)}(c)(Z-c)^i/i! \qquad \text{for all } Z \in I(Z_0). \qquad (6.19)$$

A single disc value $p_C(Z)$ is called the *(circular) centred form of p over Z*. We have the following result for $p_C(Z)$.

Theorem 6.4 Let $p(z) = \sum_{i=0}^{n} a_i z^i$ be a complex polynomial, $Z = \langle c, \rho \rangle \in I(C)$, and $p_C(Z) = \langle c_C, \rho_C \rangle$ the centred form of $p(z)$ over Z. Then

$$\left. \begin{array}{l} c_C = p(c), \\ \rho_C = \sum_{i=1}^{n} |p^{(i)}(c)| \rho^i/i! \end{array} \right\} \qquad (6.20)$$

as well as

$$\bar{p}(Z) \subseteq p_C(Z).$$

Proof From (6.11) the equations (6.20) are immediately obtained. Furthermore, equation (6.18) describes an identity for all values of $z \in C$ which implies, using inclusion msonotonicity that $\bar{p}(Z) \subseteq p_C(Z)$. □

A further comparison is possible for these inclusions of the range of $p(z)$ over Z.

Theorem 6.5 Let $p(z) = \sum_{i=0}^{n} a_i z^i$ be a complex polynomial and let $Z \in I(C)$ then it follows that

$$\bar{p}(Z) \subseteq p_C(Z) \subseteq p_H(Z) \subseteq p_S(Z). \qquad (6.21)$$

Proof If we prove that $p_C(Z) \subseteq p_H(Z)$ then the rest follows from the previous results. From (6.16) and (6.20) we have

$$c_H = c_C = p(c)$$

and

$$\rho_H = \sum_{i=1}^{n} \{(|c| + \rho)^{i-1} | \sum_{j=1}^{n} a_j c^{j-i}|\}$$

as well as

$$\rho_C = \sum_{i=1}^{n} |p^{(i)}(c)| \rho^i/i!.$$

We therefore have to prove $\rho_C \leq \rho_H$. For this we need

$$\sum_{k=i+1}^{j} \binom{k-1}{i} = \binom{j}{i+1}$$

which may be proven by complete induction. We proceed as follows (changing the initial indexing for convenience):

$$\rho_H = \rho \sum_{k=1}^{n} \{(|c| + \rho)^{k-1} | \sum_{j=k}^{n} a_j c^{j-k}|\}$$

$$= \rho \sum_{k=1}^{n} \{[\sum_{i=0}^{k-1} \binom{k-1}{i} |c|^{k-1-i} \rho^i] | \sum_{j=k}^{n} a_j c^{j-k}|\}$$

$$= \rho \sum_{i=0}^{n} \sum_{k=i+1}^{n} \{\binom{k-1}{i} |c|^{k-1-i} \rho^i | \sum_{j=k}^{n} a_j c^{j-k}|\}$$

$$= \sum_{i=0}^{n-1} \{\rho^{i+1} \sum_{k=i+1}^{n} [\binom{k-1}{i} | \sum_{j=k}^{n} a_j c^{j-(i+1)}|]\}$$

$$\geq \sum_{i=0}^{n-1} \{\rho^{i+1} | \sum_{k=i+1}^{n} [\binom{k-1}{i} \sum_{j=k}^{n} a_j c^{j-(i+1)}]|\}$$

$$= \sum_{i=0}^{n-1} \{\rho^{i+1} | \sum_{k=i+1}^{n} \sum_{j=k}^{n} \binom{k-1}{i} a_j c^{j-(i+1)}|\}$$

$$= \sum_{i=0}^{n-1} \{\rho^{i+1} | \sum_{j=i+1}^{n} \sum_{k=i+1}^{j} \binom{k-1}{i} a_j c^{j-(i+1)}|\}$$

$$= \sum_{i=0}^{n-1} \{\rho^{i+1} | \sum_{j=i+1}^{n} a_j c^{j-(i+1)} \sum_{k=i+1}^{j} \binom{k-1}{i}|\}$$

$$= \sum_{i=0}^{n-1} \{\rho^{i+1} | \sum_{j=i+1}^{n} \binom{j}{i+1} a_j c^{j-(i+1)}|\}$$

$$= \sum_{i=0}^{n-1} \{\rho^{i+1} | \frac{p^{(i+1)}(c)}{(i+1)!}|\} = \sum_{i=1}^{n} |\frac{p^{(i)}(c)}{i!}| \rho^i = \rho_C. \quad \square$$

Theorem 6.5 shows that the circular complex centred form gives an estimate of the range of a complex polynomial over a circle in the complex plane that is always smaller than that obtained through both the power-sum evaluation and the Horner-scheme.

The Circular Complex Centred Form

We now discuss the quadratic convergence of the centred form function as the radius of the domain tends to zero. In order to do this we will introduce the concept of a diameter inclusion.

Definition 6.1 Let $Z_1, Z_2 \in I(C)$. If there exists $d_1, d_2 \in Z_1 \cup Z_2$ such that $|d_1 - d_2| = 2 \operatorname{rad} Z_1$ then Z_1 is said to be *diameter included* in Z_2. This is written as

$$Z_1 \prec Z_2.$$

This definition says that $Z_1 \prec Z_2$ iff there exists two opposite points d_1, d_2 on the periphery of Z_1 which lie (together with the line segment connecting d_1 and d_2) in Z_2; or equivalently, iff one half of Z_1 lies in Z_2. Clearly, if $Z_1 \subseteq Z_2$, then $Z_1 \prec Z_2$. The converse is not true, however. The relation $\prec$ is also not transitive. Let, for example, $Z_1 = \langle 0, 2 \rangle$, $Z_2 = \langle 4, 5 \rangle$, and $Z_3 = \langle 10, 10 \rangle$. Then $Z_1 \prec Z_2$, $Z_2 \prec Z_3$, but not $Z_1 \prec Z_3$.

We will now give a brief explanation of the meanings of the diameter inclusion. First, if B is the smallest disc containing $\bar{p}(Z)$, then $B \prec B_1$ holds for any disc B_1 containing $\bar{p}(Z)$, see Lemma 6.1 in the sequel. Secondly, the diameter inclusion is the basic relation which occurs at subdivision methods for circular intervals. In the treatment of the subdivision method for rectangular intervals of I^m, the usual inclusion can be used. Considering now the circular interval $Z = \langle 0, 2 \rangle$ one may ask, how can it be subdivided? Figs 6.1 and 6.2 show the situation when Z is subdivided. Fig. 6.1 shows that one cannot find four subdiscs Z_i of Z with $\operatorname{rad} Z_i = (\operatorname{rad} Z)/2$ that cover Z. However, Fig. 6.2 shows a possible covering of Z by four subdiscs Z_i with $\operatorname{rad} Z_i = (\operatorname{rad} Z)/\sqrt{2}$. We see that $Z_i \prec Z$.

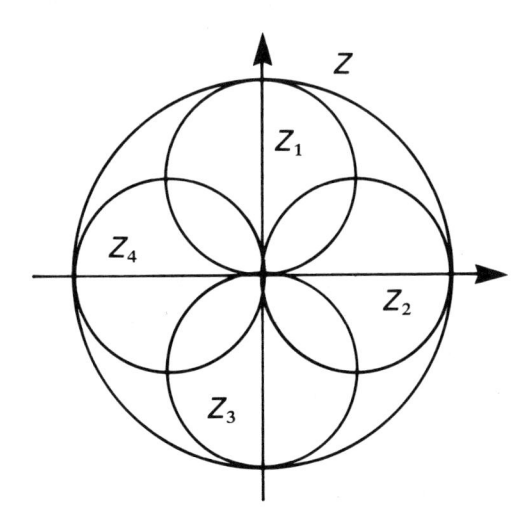

Fig. 6.1.

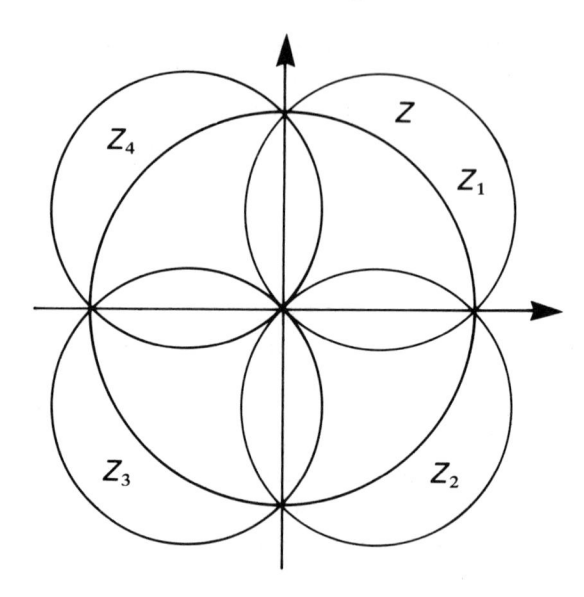

Fig. 6.2.

We now introduce the following quasimetrics q on $I(C)$, defined by

$$q(Z_1, Z_2) = |\text{rad } Z_1 - \text{rad } Z_2|$$

One can easily check that the axioms of a quasimetric (positively semidefinite, symmetry, triangle inequality) hold. (Quasimetrics are for example defined and used in Collatz (1964).)

Lemma 6.1 *Let $Z \in I(C)$ and let p be a complex polynomial. Then there exists a uniquely determined smallest disc $B \supseteq \bar{p}(Z)$ in the sense that*

$$\text{rad } B \leq \text{rad } B_1$$

for all discs $B_1 \supseteq \bar{p}(Z)$. Furthermore, $B \prec B_1$ holds for all discs $B_1 \supseteq \bar{p}(Z)$.

Proof First, we state that the lemma holds iff the lemma holds for conv $\bar{p}(Z)$ (that is the convex hull of $\bar{p}(Z)$) instead of $\bar{p}(Z)$. Since $\bar{p}(Z)$ is compact, conv $\bar{p}(Z)$ is a so-called convex body for which the lemma holds, cf. Bonnesen–Fenchel, 1934, p. 54. □

Lemma 6.2 *Let $Z = \langle c, \rho \rangle$, $Z_1 = \langle c_1, \rho_1 \rangle$, and $Z \prec Z_1$. Then*

$$|\rho - \rho_1| = q(Z, Z_1),$$
$$|c - c_1| \leq \sqrt{2\rho}\, q(Z, Z_1).$$

Proof The asserted equality is just the definition of the quasimetric q.

The inequality follows from the obvious inequality
$$\rho_1^2 \geq \rho^2 + |c - c_1|^2$$
as well as the estimation,
$$|c - c_1|^2 \leq (\rho_1 - \rho)(\rho_1 + \rho) \leq q(Z, Z_1)\, 2\rho_1. \qquad \square$$

Theorem 6.6 *Let $p(z) = \sum_{i=1}^{n} a_i z^i$ be a complex polynomial and let $Z = \langle c, \rho \rangle \in I(C)$. If $B \in I(C)$ is the smallest disc containing $\bar{p}(Z)$ then*
$$q(B, p_C(Z)) \leq \alpha \rho^2$$
where
$$\alpha = 2 \sum_{i=2}^{n} |p^{(i)}(c)| \rho^{i-2}/i!.$$

Proof Let us first construct a disc $A = \langle a, \rho_a \rangle \in I(C)$ by setting
$$a = (p(u) + p(v))/2$$
and
$$\rho_a = |p(u) - p(v)|/2$$
where
$$|p(u) - p(v)| = \max_{z_1, z_2 \in Z} \{|p(z_1) - p(z_2)|\}.$$
Then clearly
$$A \prec B.$$
For brevity we write $Z_C = p_C(Z) = \langle c_C, \rho_C \rangle$. The relation $\bar{p}(Z) \subseteq Z_C$ furthermore implies that $B \prec Z_C$. It remains to prove that
$$q(A, Z_C) \leq \alpha \rho^2.$$
From this fact together with $q(B, Z_C) \leq q(A, Z_C)$ the result then follows directly.

Let $z_1, z_2 \in Z$ such that
$$|z_1 - z_2| = 2\rho.$$
The distance between $p(z_1)$ and $p(z_2)$ can be calculated as follows:
$$|p(z_1) - p(z_2)| = \left| \sum_{i=0}^{n} \frac{p^{(i)}(c)}{i!} (z_1 - c)^i - \sum_{i=0}^{n} \frac{p^{(i)}(c)}{i!} (z_2 - c)^i \right|$$
$$= \left| p'(c)(z_1 - z_2) - \left[\sum_{i=2}^{n} \frac{p^{(i)}(c)}{i!} (z_2 - c)^i - \sum_{i=2}^{n} \frac{p^{(i)}(c)}{i!} (z_1 - c)^i \right] \right|$$

$$\geq |p'(c)| |z_1 - z_2| - |\sum_{i=2}^{n} \frac{p^{(i)}(c)}{i!} (z_2 - c)^i - \sum_{i=2}^{n} \frac{p^{(i)}(c)}{i!} (z_1 - c)^i$$

$$\geq 2\rho|p'(c)| - |\sum_{i=2}^{n} \frac{p^{(i)}(c)}{i!} (z_2 - c)^i - \sum_{i=2}^{n} \frac{p^{(i)}(c)}{i!} (z_1 - c)^i|$$

$$= 2\rho|p'(c)| - 2\sum_{i=2}^{n} \left|\frac{p^{(i)}(c)}{i!} \rho^i\right|.$$

From the definition of ρ_a it follows that

$$\rho_a \geq \tfrac{1}{2}|p(z_1) - p(z_2)|$$

$$= \rho|p'(c)| - \sum_{i=2}^{n} \left|\frac{p^{(i)}(c)}{i!}\right| \rho^i.$$

From Theorem 6.4 we have

$$\rho_C = \sum_{i=1}^{n} \left|\frac{p^{(i)}(c)}{i!}\right| \rho^i.$$

Furthermore, from the definition of q it follows that

$$q(A, Z_C) = \rho_C - \rho_a.$$

Therefore, we get

$$q(A, Z_C) \leq \left(\sum_{i=1}^{n} \left|\frac{p^{(i)}(c)}{i!}\right| \rho^i - |p'(c)|\rho + \sum_{i=2}^{n} \left|\frac{p^{(i)}(c)}{i!}\right| \rho^i \right)$$

$$= \rho^2 \left(2 \sum_{i=2}^{n} \left|\frac{p^{(i)}(c)}{i!}\right| \rho^{i-2} \right).$$

Let $\alpha = 2 \sum_{i=2}^{n} \left|\frac{p^{(i)}(c)}{i!}\right| \rho^{i-2}$.

Then we have

$$q(A, Z_C) \leq \alpha \rho_2. \qquad \square$$

Theorem 6.7 Let $Z_0 \in I(C)$ and let $p(z) = \sum_{i=0}^{n} a_i z^i$. Furthermore, let $B(Z)$ be the smallest circular interval containing $\bar{p}(Z)$ for each $Z \in I(Z_0)$. Then the centred form function $p_C(Z)$ converges quadratically to the function $B(Z)$ as rad Z tends to zero.

Proof We have to find constants α and β such that for any $Z \in I(Z_0)$,

$$|\text{rad } B(Z) - \text{rad } p_C(Z)| \leq \alpha (\text{rad } Z)^2,$$

$$|\text{mid } B(Z) - \text{mid } p_C(Z)| \leq \beta (\text{rad } Z)^2.$$

For this let $Z_0 = \langle c_0, \rho_0 \rangle$ and $Z = \langle c, \rho \rangle$. If

$$\alpha = 2 \max_{z \in Z_0} \sum_{i=2}^{n} \left| \frac{p^{(i)}(z)}{i!} \right| \rho_0^{i-2}$$

then by Theorem 6.6 and Lemma 6.1

$$\left| \operatorname{rad} B(Z) - \operatorname{rad} p_C(Z) \right| = q(B(Z), p_C(Z)),$$

$$\rho^2 \sum_{i=2}^{n} \left| \frac{p^{(i)}(z)}{i!} \right| \rho^{i-2} \leq \rho^2 \alpha.$$

Setting

$$\gamma = \max_{z \in Z_0} \sum_{i=1}^{n} \left| \frac{p^{(i)}(z)}{i!} \right| \rho_0^i$$

which is an upper bound for the radius of $p_C(Z)$ that is independent of Z, we get by Theorem 6.6 and Lemma 6.2 that

$$\left| \operatorname{mid} B(Z) - \operatorname{mid} p_C(Z) \right| \leq \sqrt{2} \operatorname{rad} p_C(Z) \, q(B(Z), p_C(Z))$$
$$\leq \alpha \sqrt{2\gamma} \rho^2.$$

Hence, $\beta = \alpha \sqrt{2\gamma}$ is a constant for the second inequality as desired. □

Remark 6.2 Theorem 6.7 is also valid if the condition $Z \subseteq Z_0$ for the circular interval variable is replaced by

$$Z \prec Z_0.$$

This weakening is of importance if the subdivision method is used as explained by Fig. 6.2.

Example 6.3 This example is typical of the inclusions that are obtained for the range of a complex polynomial using the methods discussed in this section. The polynomial

$$\begin{aligned}
p(z) = {} & (0.112 - i0.1022) + (0.1056 + i0.1096)z \\
& - (0.1036 + i0.1219)z^2 + (0.1037 + i0.219)z^3 \\
& + (0.1058 - i0.1921)z^4 + (0.1072 - i0.1921)z^5 \\
& + (0.1091 - i0.1039)z^6 + (0.1039 - i0.2018)z^7 \\
& + (0.1652 - i0.1036)z^8 + (0.1036 - i0.1005)z^9
\end{aligned}$$

is to be evaluated over the disc $Z = \langle -0.1023 + i0.2011, 0.9913 \rangle$. Using the three methods discussed earlier the following inclusion estimates for the range $\bar{p}(Z)$ are obtained.

Evaluation	Resulting circle	Area of resulting circle
Power sum	$(0.1224 - i0.0591, 4.354)$	59.55
Horner scheme	$(0.1224 - i0.0591, 4.069)$	52.01
Centred form	$(0.1224 - i0.0591, 1.910)$	11.46

These circles are plotted in Fig. 6.3. Furthermore, by computing the value of the polynomial at 200 uniformly spaced points along the domain circle and then joining together the values obtained a close polygonal approximation to the range is plotted.

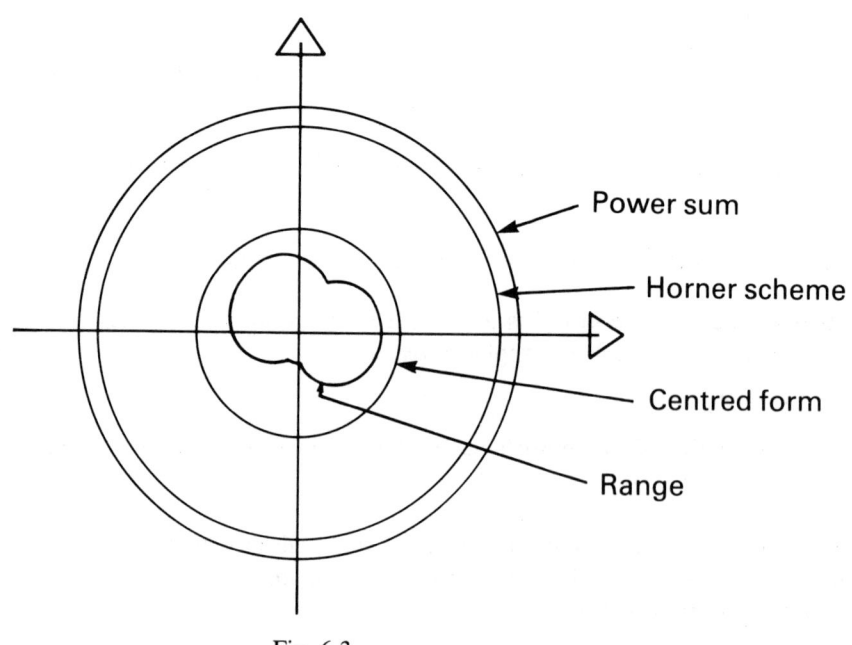

Fig. 6.3.

Example 6.4 A further example is given by the polynomial

$$p(z) = (0.15 - i0.1) + (0.15 - i0.12)z + (-0.2 - i0.2)z^2$$
$$+ (0.1 + i0.3)z^3 + (0.1 - i0.2)z^4 + (0.1 - i0.2)z^5$$
$$+ (0.2 - i0.2)z^6 + (0.1 - i0.2)z^7 + (0.2 - i0.1)z^8$$
$$+ (0.1 - i0.1)z^9$$

evaluated over the disc $Z = \langle-0.1 + i0.2, 0.9\rangle$. The results are given below. A plot is given in Fig. 6.4.

Evaluation	Resulting circle	Area of resulting circle
Power sum	$\langle 0.1590 - i0.0405, 3.221\rangle$	37.58
Horner scheme	$\langle 0.1590 - i0.0405, 3.059\rangle$	29.40
Centred form	$\langle 0.1590 - i0.0405, 1.718\rangle$	9.272

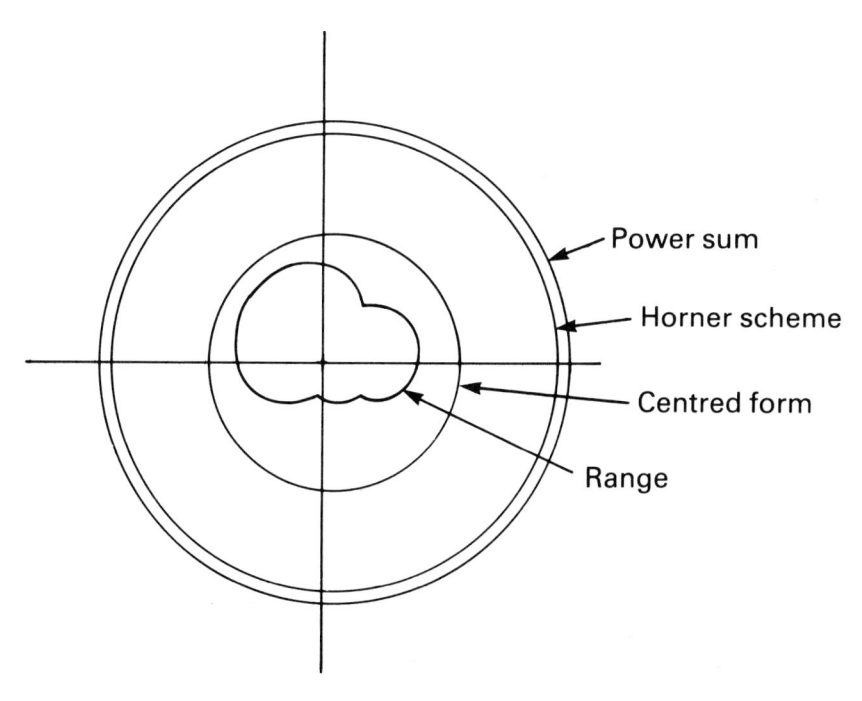

Fig. 6.4.

Example 6.5 The polynomial

$$p(z) = (-0.1 - i0.1) + (-i0.1)z$$
$$+ (-0.1 - i0.1)z^2 + (0.4 + i0.3)z^3$$
$$+ (-0.1 - i0.1)z^4 + (0.1 - i0.1)(z^5 + z^6 + z^7)$$

128   Other Inclusions for the Range of a Function   [Ch. 6

is to be evaluated over the circle $Z = \langle 1.0 - i0.5, 1.0 \rangle$. The results are given below.

Evaluation	Resulting circle	Area of resulting circle
Power sum	$\langle -0.6117 - i0.1117, 55.40 \rangle$	9636.0
Horner scheme	$\langle -0.6117 - i0.1117, 42.17 \rangle$	5586.0
Centred form	$\langle -0.6117 - i0.1117, 41.54 \rangle$	5419.0

A plot is given in Fig. 6.5. Changing the domain to $Z = \langle -0.40 + i0.60, 1.0 \rangle$ gives the results below. A plot is given in Fig. 6.6.

Evaluation	Resulting circle	Area of resulting circle
Power sum	$\langle 0.08846 + i0.1381, 14.28 \rangle$	640.4
Horner scheme	$\langle 0.08846 + i0.1381, 9.424 \rangle$	279.0
Centred form	$\langle 0.08846 + i0.1381, 6.923 \rangle$	150.6

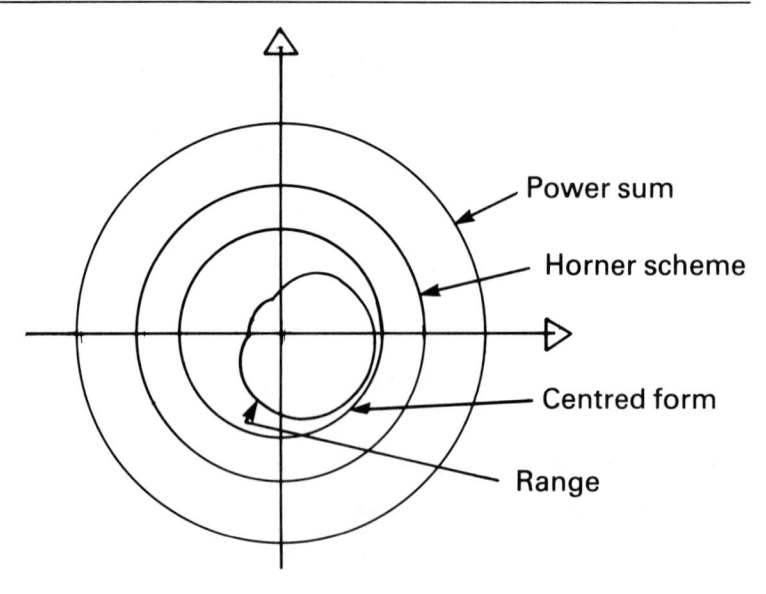

Fig. 6.5.

The Circular Complex Centred Form

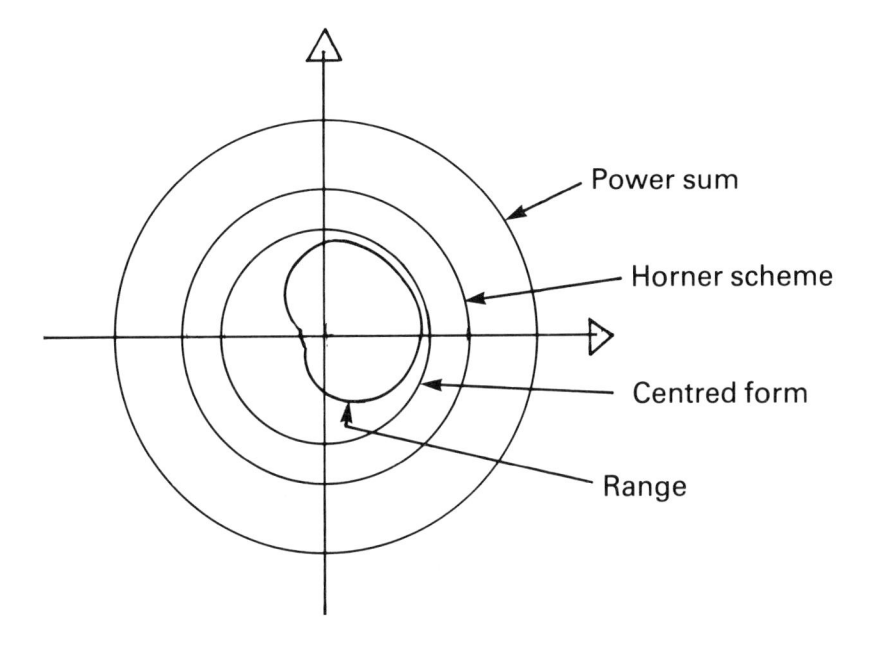

Fig. 6.6.

Finally we want to describe the circular centred form for complex rational functions over C^m, which was developed by Krawczyk (1983). We can be brief since this form is only the analogy of the real case developed in Section 2.6, for the complex case. For handling rational functions, the circular complex arithmetic (6.11) is to be supplemented by division in $I(C)$. If $Z = \langle c, \rho \rangle$ and $Z_1 \in I(C)$, then we define

$$1/Z = \langle \bar{c}/(|c|^2 - \rho^2), \rho/(|c|^2 - \rho^2) \rangle \quad \text{if } \rho < |c|,$$
$$Z_1/Z = Z_1(1/Z) \quad \text{if } \rho < |c|.$$

Now we only have to extend the concepts of a function procedure of the corresponding function, interval function procedure, interval function, and of an interval slope to the complex case. This means, that the constants $b_1,\ldots,b_r$ and the variables $x_1,\ldots,X_m$, $u_1,\ldots,u_s$ in Definition 2.4 are allowed to be constants or variables of C. Let $e_i^c \in I(C)^m$ for $i = 1,\ldots,m$ denote the point vector, the jth component of which is $\langle 1, 0 \rangle$ if $j = i$ and $\langle 0, 0 \rangle$ if $j \neq i$, and let $o^c \in I(C)^m$ denote the vector $(\langle 0, 0 \rangle,\ldots,\langle 0, 0 \rangle)$. Let now a complex function procedure S be given and $X \in I(C)^m$. Then the definition of the complex interval slope of S over X arises from Definition 2.5 if in Definition 2.5 the vectors e_i and o are replaced by the vectors e_i^c and o^c and the arithmetic is replaced by the arithmetic for circular intervals. Then, like the real case, $G \in I(C)^m$ is an interval slope of f over X if f is the function defined by the complex function procedure S and G is the complex interval slope of S over X.

The circular interval

$$F(X) = f(c) + G \cdot (X - c)$$

where c is the midpoint of X or any other point vector of X, is called *Krawczyk's circular centred form*. Clearly, $F(X) \supseteq \tilde{f}(X)$. The computational complexity is again very low, i.e. if k arithmetic operations are needed for describing the procedure S, then $O(k)$ circular interval operations are needed for the computation of the form $F(X)$.

6.3 THE METHODS OF CARGO–SHISHA AND RIVLIN

An interesting method for bounding the range of a polynomial over an interval is obtained through the use of Bernstein forms. These forms are intimately connected to Bernstein polynomials. A survey of Bernstein polynomials and some of their applications is found in Lorenz (1953). The first application to the range of polynomials was given by Cargo–Shisha (1966). Improvements and convergence results were given by Rivlin (1970) and algorithms and applications to interval polynomials by Rokne (1977, 1981).

In this section we discuss the ideas of Cargo–Shisha and Rivlin and we show how these forms may be used to obtain bounds for the range of a polynomial over an interval. We also prove that the Bernstein form converges linearly to the range of the polynomials with respect to the order of the forms. The slow convergence is compensated for by obtaining criteria that indicate whether the calculated estimation is the range or not.

We also discuss a further method by Cargo–Shisha (1966) and Rivlin (1970) based on a simple estimation of the second derivative in a Taylor expansion of a polynomial.

In the following we develop estimates for a polynomial $p(x) = \sum_{i=0}^{n} a_i x^i$ over the interval $[0, 1]$. This is no restriction since estimates for the values of the polynomial over $X = [a, b]$ are obtained by a linear transformation of the domain X to $[0, 1]$ which leaves the range $\bar{p}(X)$ invariant.

We first introduce the Bernstein functions.

Definition 6.2 For $k \geq 0$ the *Bernstein functions* B_j^k are defined to be

$$B_j^k(x) = \binom{k}{j} x^j (1 - x)^{k-j} \quad j = 0, 1, 2, \ldots, k$$

where $0^0 = 1$.

The following two lemmas, due to Rivlin (1970) describe some important properties of the Bernstein functions.

The Methods of Cargo-Shisha and Rivlin

Lemma 6.3 *Let $s, k \in \mathbb{N}$ such that $k \geq s \geq 0$. It then follows that*

$$x^s = \sum_{j=s}^{k} \frac{\binom{j}{s}}{\binom{k}{s}} B_j^k(x).$$

Proof We calculate

$$x^s = x^s(x + (1-x))^{k-s} = \sum_{i=0}^{k-s} \binom{k-s}{i} x^{s+i}(1-x)^{k-s-i}$$

$$= \sum_{j=s}^{k} \frac{\binom{j}{s}}{\binom{k}{s}} \binom{k}{j} x^j (1-x)^{k-j} = \sum_{j=s}^{k} \frac{\binom{j}{s}}{\binom{k}{s}} B_j^k(x). \qquad \square$$

Based on this lemma we prove that a polynomial may be represented uniquely in terms of Bernstein functions.

Lemma 6.4 *Let $p(x)$ be a polynomial of degree n. If $k \geq n$ then p is a unique linear combination*

$$p(x) = \sum_{j=0}^{k} b_j^k B_j^k(x) \qquad (6.22)$$

of $B_j^k(x)$, $j = 0, \ldots, k$.

Proof Using the previous lemma we obtain

$$p(x) = \sum_{i=0}^{n} a_i x^i = \sum_{i=0}^{n} a_i \sum_{j=i}^{k} \frac{\binom{j}{i}}{\binom{k}{i}} B_j^k(x)$$

$$= \sum_{j=0}^{k} B_j^k(x) \sum_{i=0}^{j} a_i \frac{\binom{j}{i}}{\binom{k}{i}} = \sum_{j=0}^{k} b_j^k B_j^k(x)$$

where

$$b_j^k = \sum_{i=0}^{j} a_i \frac{\binom{j}{i}}{\binom{k}{i}} \qquad j = 0, \ldots, k. \qquad (6.23)$$

The representation (6.22) says that the polynomials $B_0^k, \ldots, B_k^k$ span the space of all polynomials of degree smaller than or equal to k. Thus, they form a basis of this space which gives the asserted uniqueness. $\square$

The unique representation of a polynomial p of degree n over $[0, 1]$ in the form

$$p(x) = \sum_{j=0}^{k} b_j^k B_j^k(x)$$

for a $k \geq n$ is called the *Bernstein form of order k* for p. The coefficients b_j^k are called the *Bernstein coefficients* of p (of order k).

The following theorem, essentially due to Cargo–Shisha (1966), shows that the Bernstein coefficients provide bounds for the range of the polynomial.

Theorem 6.8 Let $k, n \in N$ with $k \geq n \geq 0$ and let

$$p(x) = \sum_{j=0}^{k} b_j^k B_j^k(x)$$

be the Bernstein form of p of order k. Then

$$\bar{p}([0, 1]) \subseteq \left[\min_{0 \leq j \leq k} b_j^k, \max_{0 \leq j \leq k} b_j^k\right].$$

Proof By Lemma 6.4 we get

$$p(x) = \sum_{j=0}^{k} b_j^k B_j^k(x).$$

From Lemma 6.3 it follows that

$$1 = x^0 = \sum_{j=0}^{k} \binom{k}{j} x^j (1-x)^{k-j} = \sum_{j=0}^{k} B_j^k(x).$$

Therefore for each $x \in [0, 1]$ the sum $\sum_{j=0}^{k} b_j^k B_j^k(x)$ is a convex combination of the Bernstein coefficients b_j^k, $j = 0, 1, \ldots, k$ of order k. This means that for all $x \in [0, 1]$ we get

$$\min_{0 \leq j \leq k} b_j^k \leq \sum_{j=0}^{k} b_j^k B_j^k(x) \leq \max_{0 \leq j \leq k} b_j^k$$

or equivalently.

$$\bar{p}([0, 1]) \subseteq \left[\min_{0 \leq j \leq k} b_j^k, \max_{0 \leq j \leq k} b_j^k\right]. \qquad \square$$

The next theorem also due to Cargo–Shisha (1966) is concerned with an important property of the Bernstein form, namely when the estimate provided by the Bernstein form is the range.

Theorem 6.9 Let $k, n \in N$ with $k \geq n \geq 0$ and let $p(x) = \sum_{j=0}^{k} b_j^k B_j^k(x)$ be the Bernstein form for p of order k and let $\bar{p}([0, 1]) = [a, b]$. Then

$$a = \min_{0 \leq j \leq k} b_j^k \text{ iff } \min_{0 \leq j \leq k} b_j^k = \min(b_0^k, b_k^k)$$

and

$$b = \max_{0 \leq j \leq k} b_j^k \text{ iff } \max_{0 \leq j \leq k} b_j^k = \max(b_0^k, b_k^k).$$

Proof We first note that

$$b_0^k = \sum_{i=0}^{0} a_i \frac{\binom{i}{i}}{\binom{0}{i}} = a_0 = p(0)$$

and

Sec. 6.3]   The Methods of Cargo-Shisha and Rivlin   133

$$b_k^k = \sum_{i=0}^{k} a_i \frac{\binom{k}{i}}{\binom{k}{i}} = \sum_{i=0}^{k} a_i = p(1).$$

Suppose now that $a = \min_{0 \leq j \leq k} b_j^k$ and that $x \in (0, 1)$. Then if $b_0^k = b_1^k = \ldots = b_k^k$ we have that $p(0) = b_0^k = a = b_k^k = p(1)$ and the theorem is valid. If $b_0^k = b_1^k = \ldots = b_k^k$ is not satisfied then $\sum_{j=0}^{k} b_j^k \binom{k}{j} x^j (1-x)^{k-j} > \min_{0 \leq j \leq k} b_j^k$ which means that the minimum cannot occur at an interior point of $[0, 1]$. Hence it must ocurr at an endpoint of the interval.

Conversely, suppose that $\min_{0 \leq j \leq k} b_j^k = \min(b_0^k, b_k^k)$ and assume that $\min_{0 \leq j \leq k} = b_0^k$ for simplicity. Then the bound is sharp, i.e. $a = p(0) = b_0^k = \min_{0 \leq j \leq k} b_j^k$. The same argument is valid if $\min_{0 \leq j \leq k} b_j^k = b_k^k$.

The second part of the theorem regarding the sharpness of the upper bound is proven in the same manner. □

An important consequence of Theorem 6.9 is given in the following corollary.

Corollary 6.1 *Let $p(x)$ of degree $n \geq 0$ be monotonic over $[0, 1]$. Then for all $k \geq n$*

$$\bar{p}([0, 1]) = \left[\min_{0 \leq j \leq k} b_j^k, \max_{0 \leq j \leq k} b_j^k \right],$$

that is the Bernstein form of order k computes the range.

Proof The corollary is an immediate consequence of Theorem 6.9. □

Based on the above results we now make the following definition.

Definition 6.3 *Let $p(x)$ be a polynomial of degree $n \geq 0$ and let $k \geq n$. Then the interval*

$$B_k = \left[\min_{0 \leq j \leq k} b_j^k, \max_{0 \leq j \leq k} b_j^k \right]$$

is called the Bernstein approximation of order k to $\bar{p}([0, 1])$.

We now present results due to Rivlin (1970) which show that the Bernstein approximations B_k converge at least linearly to $\bar{p}([0, 1])$. For this we need an idea from approximation theory due to Bernstein (1912) given in the following definition.

Definition 6.4 *Let $f: [0, 1] \to R$ be a real function. The Bernstein polynomial of f of order k is then defined by*

$$B_k(f, x) = \sum_{j=0}^{k} f\left(\frac{j}{k}\right) B_j^k(x).$$

If now the function f in Definition 6.4 is the polynomial x^s, $0 \leq s \leq k$, then

$$B_k(x^s, x) = \sum_{j=0}^{k} \left(\frac{j}{k}\right)^s B_j^k(x)$$

and from this it follows that

$$B_k(x^s, x) - x^s = \sum_{j=0}^{s-1} \left(\frac{j}{k}\right)^s B_j^k(x) + \sum_{j=s}^{k} \left|\left(\frac{j}{k}\right)^s - \frac{\binom{j}{s}}{\binom{k}{j}}\right| B_j^k(x)$$

$$= \sum_{j=0}^{k} \delta_j(s) B_j^k(x) \tag{6.24}$$

using Lemma 6.3 and defining

$$\delta_j(s) = \begin{cases} \left(\dfrac{j}{k}\right)^s & \text{if } j < s \\ \left(\dfrac{j}{k}\right)^s - \dfrac{\binom{j}{s}}{\binom{k}{j}} & \text{if } s \leq j \leq k. \end{cases}$$

The factors $\delta_j(s)$ may now be estimated as in the following theorem.

Theorem 6.10 *If* $k \geq n > 1$ *then*

$$\delta_j(s) \leq \frac{(s-1)^2}{k} \quad \text{for } j = 0,1,\ldots,k, \ s = 0,1,\ldots,n.$$

Proof It is known that $B_k(1,x) = 1$ and $B_k(x,x) = x$, see for example, Cheney (1966). This means that $\delta_j(0) = \delta_j(1) = 0$, $j = 0,1,\ldots,k$. We may therefore assume $s \geq 2$.

From (6.24) we have for $0 \leq j < s$ that

$$\delta_j(s) = \left(\frac{j}{k}\right)^s \leq \left(\frac{s-1}{k}\right)^s \leq \left(\frac{s-1}{k}\right)^s \leq \left(\frac{s-1}{k}\right)^s.$$

We now consider $2 \leq s \leq j$. Then

$$\left(\frac{j}{k}\right)^s - \frac{\binom{j}{s}}{\binom{k}{j}} = \left(\frac{j}{k}\right)^s - \frac{j!(k-s)!}{(j-s)!k!}$$

$$= \left(\frac{j}{k}\right)^s - \frac{j(j-1)\cdots(j-(s-1))}{k(k-1)\cdots(k-(s-1))}$$

$$= \left(\frac{j}{k}\right)^s \left[1 - \frac{(1-1/j)\cdots(1-(s-1)/j)}{(1-1/k)\cdots(1-(s-1)/j)}\right]$$

$$\leq \left(\frac{j}{k}\right)^s \left[1 - \left(1 - \frac{1}{j}\right)\cdots\left(1 - \frac{s-1}{j}\right)\right]$$

$$\leq \left(\frac{j}{k}\right)^s \left[1 - \left(1 - \frac{s-1}{j}\right)^{s-1}\right].$$

Applying the mean-value theorem to $(1 - x)^{s-1}$ we obtain

$$1 - \left(1 - \frac{s-1}{j}\right)^{s-1} \leq \frac{(s-1)^2}{j}$$

from which

$$\delta_j(s) \leq \frac{(s-1)^2}{k} \left(\frac{j}{k}\right)^{s-1} \leq \frac{(s-1)^2}{k}$$

follows. □

Based on the above results we now show as in Rivlin (1970) that the Bernstein approximations converge to the range and that the convergence is at least linear in the order of the approximations.

Theorem 6.11 *Let $p(x)$ be a polynomial of a degree $n > 0$ and $k \geq n$. Then*

$$w\left(\left[\min_{0 \leq j \leq k} b_j^k, \max_{0 \leq j \leq k} b_j^k\right]\right) - w(\bar{p}[0, 1]) \leq A/k$$

where $A = 2 \sum_{s=2}^{n} (s-1)^2 |p^{(s)}(0)|/s!$.

Proof From (6.24) we obtain

$$B_k(p,x) - p(x) = \sum_{s=0}^{n} a_s[B_k(x^s, x) - x^s]$$

$$= \sum_{s=0}^{n} a_s \sum_{j=0}^{k} \delta_j(s) B_j^k(x)$$

where $a_s = p^{(s)}(0)/s!$. Writing

$$B_k(p,x) - p(x) = \sum_{j=0}^{k} \delta_j B_k^j(x)$$

which by using Theorem 6.10 results in

$$\delta_j = \sum_{s=2}^{n} a_s \delta_j(s) \leq \frac{1}{k} \sum_{s=2}^{n} \frac{(s-1)^2}{s!} |p^{(s)}(0)|$$

from which $|\delta_j| \leq A/2k$ follows. Since this is the excess width on each end the bound is obtained. □

Remark 6.3 In Theorem 6.11 a bound is given that depends on the order of the Bernstein approximation, but not on the width of the domain which is fixed at $[0, 1]$.

Remark 6.4 The Bernstein coefficients b_j^k may be computed by difference tables, see Cargo–Shisha (1966) and Rokne (1979a).

In Cargo–Shisha (1966) a further method is suggested for the inclusion of the range of values of a polynomial p over $X \in I$. Although we describe the method relative to $[0, 1]$, the application relative to any $X \in I$ is self-evident. The method is based on the computation of function values at a set of points in $[0, 1]$. The following theorem by Rivlin (1970) is an improvement on the result of Cargo–Shisha (1966).

Theorem 6.12 *Let $k > 0, 0 = t_0 < t_1 < \ldots < t_k = 1$ and let $d_k = \max(t_{j+1} - t_j), j = 0, 1, \ldots, k - 1$. Then*

$$\min_{0 \le j \le k} p(t_j) - \frac{d_k^2}{8} \max_{x \in [0,1]} |p^2(x)| \le p(x) \le \max_{0 \le j \le k} p(t_j) + \frac{d_k^2}{8} \max_{x \in [0,1]} |p^2(x)|. \tag{6.25}$$

Proof Let $\bar{p}([0, 1]) = [a, b]$. Then there exists a $\xi \in [0, 1]$ such that $p(\xi) = b$. Furthermore, for some $0 \le j \le k$ we have

$$|t_j - \xi| \le |t_i - \xi|, \qquad i = 0, 1, \ldots, k$$

as well as

$$|t_j - \xi| \le d_k/2.$$

From Taylor's theorem we obtain

$$p(t_j) = p(\xi) + (t_j - \xi) p'(\xi) + (t_j - \xi)^2 p''(\eta)/2 \tag{6.26}$$

where $\eta \in [0, 1]$. If $\xi = 0$ or 1 then the right-most inequality in (6.25) is trivially true. If $0 < \xi < 1$ then $p'(\xi) = 0$ and (6.26) implies that

$$b \le \max_{0 \le j \le k} p(t_j) + (t_j - \xi)^2 \max_{x \in [0,1]} |p''(x)|/2$$

$$\le \max_{0 \le j \le k} p(t_j) + d_k^2 \max_{x \in [0,1]} |p''(x)|/8.$$

The lower bound is established analogously. □

Corollary 6.2 *Let $k > 0$ and set $r_k = \dfrac{1}{8k^2} \sum_{j=0}^{n} (j - 1)j|a_j|$. Then*

$$\bar{p}([0, 1]) \subseteq \left[\min_{0 \le j \le k} p\!\left(\frac{j}{k}\right) - r_k, \ \max_{0 \le j \le k} p\!\left(\frac{j}{k}\right) + r_k \right].$$

Proof Setting $t_i = i/k$, $i = 0, 1, \ldots, k$ in Theorem 6.12 it follows that $d_k = 1/k$. For each $x \in [0, 1]$ it furthermore follows that

$$|p''(x)| = \left| \sum_{j=0}^{n} j(j-1) a_j \right| \le \sum_{j=0}^{n} j(j-1) |a_j|$$

and the result is evident. □

6.4 THE METHOD OF DUSSEL–SCHMITT FOR POLYNOMIALS

A further method for computing the range of a real polynomial over a real interval is given in Dussel–Schmitt (1970). The principle of the method is simple. A polynomial p of degree n has at most $n + 1$ relative minima and maxima on a compact interval. These relative extremal values are computed via the zeros of p'. This is the basic mathematical idea.

Numerically, further complications arise:
(i) the search for all the zeros of p' is involved and unstable,
(ii) the numerical computations only give points 'near' the zeros and therefore only points 'near' the relative extremal values of p.

The method of Dussel–Schmitt uses knowledge about monotonicity of the derivatives of p to overcome these difficulties. The search for the zeros is facilitated and the computation is controlled in such a manner that the inclusion of the range is guaranteed, contrary to what one would expect from (ii) above.

The description of the method now follows that of Dussel–Schmitt closely, only omitting some inessential details.

Let $p(x) = \sum_{i=0}^{n} a_i x^i$ be a polynomial of exact degree n. Furthermore let $X \in I$. We are required to compute a good outer approximation $P(X)$ to $\bar{p}(X)$. In the sequel if $Y \in I(X)$ then the notation $p^{(i)}(Y)$ is used to denote the natural interval extension approximation to $\bar{p}^{(i)}(Y)$. This approximation may be computed using, for example, the Horner-scheme, or the standard centred form. The approach selected by Dussel–Schmitt (1970) consists of two stages, a differentiation stage followed by an integration stage. In the differentiation stage an integer i_0 is determined such that $0 \notin p^{(i_0)}(X)$.

Clearly $i_0 \leq n$ since $p^{(n)}(x) = a_n$, a constant $\neq 0$ by assumption. In the integration stage inclusions for the zeros of $p^{(i_0 - j)}$, $j = 1, 2, \ldots, i_0 - 1$ are determined in that sequence. The advantage of proceeding in this manner lies in the fact that not only are inclusions for the zeros of p' computed, but information about the slopes of p' in the intervals containing the zeros is obtained. The inclusions for the zeros of p' on X then provide intervals where the relative extremal values of p are found. Evaluating p over these intervals using the additional information on the slopes of p' will therefore result in lower and upper bounds for the range of p over X.

The implementation of the differentiation stage poses no problems algorithmically or numerically. The detailed implementation of the integration stage is, however, more difficult and we therefore proceed with a fairly complete description of the algorithmic implementation of this stage, also considering the numerical problems that arise due to the use of finite length floating point arithmetic.

The first two steps of the integration stage are now described. It was shown in the differentiation stage that $p^{(i_0)}(x) \neq 0$ for $x \in X$. This means that $p^{(i_0-1)}$ is strictly monotonic in X. Therefore it follows that $p^{(i_0-1)}$ has at most one zero in X. This zero is enclosed in an interval $[y_0, x_1]$ using, for example, the procedure developed by Nickel (1967). The polynomial $p^{(i_0-2)}$ is now strictly monotonic both in $[a, y_0]$ and in $[x_1, b]$ and it has therefore at most one zero in each of these intervals. It is, however, not possible to make any statement about the monotonicity of $p^{(i_0-2)}$ in the interval $[y_0, x_1]$. If therefore $0 \in p^{(i_0-2)}([y_0, x_1])$ then a special computation, to be described below, must be performed in order to decide whether a subinterval of $[y_0, x_1]$ must be considered to contain a zero of $p^{(i_0-2)}$. For example this is the case if $|p^{(i_0-2)}(x)| \leq \delta$, where δ is the smallest machine respresentable floating point number, for some $x \in [y_0, x_1]$. A second possibility will be treated later.

In the general case we consider $p^{(i)}$ of degree $n - i$. The zeros of $p^{(i)}$ are enclosed in m intervals $[y_l, x_{l+1}]$, $l = 0, 1,..., m - 1$, where $x_l < y_l$, $l = 1, 2,..., m - 1$. These intervals are defined using the condition $|p^{(i)}(x)| < \delta$ for some $x \in [y_l, x_{l+1}]$. The number of intervals m may be larger than the number of zeros of $p^{(i)}$. We therefore call these intervals *suspect intervals* since it is only suspected that they contain a zero of $p^{(i)}$. Furthermore, bounds q_{l+1} and z_{l+1} are also given for $\bar{p}^{(i)}([y_l, x_{l+1}])$ and therefore for the slope of $p^{(i-1)}$ in $[y_l, x_{l+1}]$ for $l = 0, 1,..., m - 1$. These bounds satisfy $q_{l+1} < 0 < z_{l+1}$, $l = 0, 1,..., m - 1$. We seek the suspect intervals of $p^{(i-1)}$ as well as the bounds for the slope of $p^{(i-1)}$ over these intervals.

The polynomial $p^{(i-1)}$ is now strictly monotonic in the intervals $[x_l, y_l]$, $l = 0, 1,..., m$ ($x_0 = a$, $y_m = b$). In the intervals $[y_l, x_{l+1}]$, $l = 0, 1,..., m - 1$ we only have bounds q_{l+1} and z_{l+1} for the slope of $p^{(i-1)}$. We therefore call $[x_l, y_l]$ a *monotonicity interval* and $[y_l, x_{l+1}]$ a *non-monotonicity interval*. In order to determine the non-monotonicity intervals $[v_k, w_{k+1}]$ of $p^{(i-1)}$ as well as the estimate r_{k+1} and s_{k+1} for the slope of $p^{(i-2)}$ in $[v_k, w_{k+1}]$ we now investigate pairwise the monotonicity interval $[x_l, y_l]$ and the non-monotonicity interval $[y_l, x_{l+1}]$. The details of this investigation are now dealt with. The signs of $p^{(i-1)}(x_l)$ and $p^{(i-1)}(y_l)$ are compared in order to decide whether or not $p^{(i-1)}$ has a zero in the monotonicity interval $[x_l, y_l]$. If it has no zero in this interval then one proceeds at once to the non-monotonicity interval $[y_l, x_{l+1}]$. The same happens when the interval $[x_l, y_l]$ has zero width (this can only occur when $l = 0$, that is $[a, y_0]$, or $l = m$, that is $[x_m, b]$). If $p^{(i-1)}$ does not have a zero in $[x_l, y_l]$, then it is included in an interval $[u_0, u_1]$ using the zero-finding procedure. At the same time bounds r and s are computed for the slope of $p^{(i-2)}(x)$ in $[u_0, u_1]$. If the interval $[u_0, u_1]$ borders on a previously calculated non-monotonicity interval, $[v_k, w_{k+1}]$ with the bounds r_{k+1} and s_{k+1} then the intervals $[v_k, w_{k+1}]$ and $[u_0, u_1]$ are joined together forming an interval $[v_k, w_{k+1}] = [v_k, u_1]$. The new bounds for the slope are taken as the minimum of r_{k+1} and r as well as the maximum of s_{k+1} and s.

Sec. 6.4] The Method of Dussel-Schmitt for Polynomials 139

If the two zeros do not border on each other then we set $[v_{k+1}, w_{k+2}] := [u_0, u_1]$, $r_{k+2} = r$, $s_{k+2} = s$. Furthermore, the index k is increased by one.

One now proceeds to the non-monotonicity interval $[y_l, x_{l+1}]$. Since one is not able to make any conclusions about the monotonicity one starts with the question of whether

$$0 \in p^{(l-1)}([y_l, x_{l+1}])$$

in order to determine whether or not one should suspect the interval $[y_l, x_{l+1}]$ of having a zero. If it turns out that $0 \in p^{(l-1)}([y_l, x_{l+1}])$ then one should set $[v_k, w_{k+1}] := [y_l, x_{l+1}]$ and therefore consider $[y_l, x_{l+1}]$ to be an interval suspected of having a zero. It turns out that this gives too coarse results when $[y_l, x_{l+1}]$ is a zero of multiplicity greater than one. At this point a procedure that treats double zeros is used. This procedure will be described in Remark 6.5 at the end of this section.

Now it suffices to note that a linear inclusion for $p^{(i-1)}$ in $[y_l, x_{l+1}]$ using the estimates q_{l+1} and z_{l+1} for the slope of $p^{(i-1)}$, is computed. Using this linear inclusion a smaller interval $[u_0, u_1]$ as well as smaller bounds r and s are computed for the slope of $p^{(i-2)}$ in $[u_0, u_1]$. Depending on the case this new interval is either joined with an existing bordering interval or it is treated as a new non-monotonicity interval. One then proceeds to the next pair $[x_{l+1}, y_{l+1}]$, $[y_{l+1}, x_{l+2}]$.

When all non-monotonicity intervals $[v_k, w_{k+1}]$ for $p^{(i-1)}$ as well as the corresponding bounds r_{k+1} and s_{k+1} for the slopes of $p^{(i-2)}$ have been determined then one proceeds to determine the zeros of $p^{(i-2)}$ in the same manner as above obtaining k_{i-2} zeros. From the above integration phase we now have intervals $[v_l, w_{l+1}]$, $l = 0,1,\ldots, k_1 - 1$ containing the zeros of p' in $[a, b]$. In order to compute an estimate $P(X)$ for the range $\bar{p}(X)$, one might then simply compute the reals or the intervals $P_0 = p(a)$, $P_l = p([v_l, w_{l+1}])$, $l = 0, 1,\ldots, k_1 - 1$ and $P_{k_1} = p(b)$ and then set

$$P(X) = \left[\min_{0 \leq l \leq k_1} (\inf P_l), \max_{0 \leq l \leq k_1} (\sup P_l) \right].$$

If the width of the interval $[v_l, w_{l+1}]$ is comparatively large then the evaluation $p([v_l, w_{l+1}])$ will in general give poor bounds using any of the previously mentioned methods. One therefore tries to improve these bounds by enclosing p in a rhomboid in each interval $[v_l, w_{l+1}]$, $l = 0,1,\ldots,k_1 - 1$. The sides of this rhomboid are constructed using the values $p(v_l)$ and $p(w_{l+1})$ as well as the slopes r_{l+1} and s_{l+1} as calculated in the last integration step. This results in

g_1 being the line through the point $(v_l, p(v_l))$ with the slope s_{l+1}.
g_2 being the line through the point $(v_l, p(v_l))$ with the slope r_{l+1},
g_3 being the line through the point $(w_{l+1}, p(w_{l+l}))$ with the slope r_{l+1},
g_4 being the line through the point $(w_{l+1}, p(w_{l+l}))$ with the slope s_{l+1}.

The intersection of the lines g_1 and g_3 is denoted by $S_1 = (c_1, t_1)$ and the intersection of the lines g_2 and g_4 by $S_2 = (c_2, t_2)$. We then have $\tilde{P}_l = [t_1, t_2] \supseteq \bar{p}([v_l, w_{l+1}])$. An improvement is obtained by taking the intersection of $\tilde{P}_l$ and P_l obtaining

$$P_l^* := P_l \cap \tilde{P}_l \supseteq \bar{p}([v_l, w_{l+1}]).$$

In this manner a sequence $P_0^* := P_0, P_1^*, \ldots, P_{k_1-1}^*, P_{k_1}^* := P_{k_1}$ is computed from which

$$P(X) := \left[\min_{0 \leq l \leq k} (\inf P_l^*), \max_{0 \leq l \leq k} (\sup P_l^*) \right]$$

is computed.

This estimate $P(X)$ of the range $\bar{p}(X)$ converges to the range for increasing word-length of the computing device employed, provided the zero-finding procedure also finds the zeros with increasing accuracy (in the limit a point).

Remark 6.5 Particular considerations have to be taken care of in the case of a double zero since the zero-finding procedure tends to produce poor results for double zeros. Let $p^{(i)}$ satisfy the following conditions in $Y = [c, d]$,

(a) $0 \in p^{(i)}(Y)$

(b) $q \leq p^{(i+1)}(x) \leq z$ for $x \in Y$.

Then $p^{(i)}$ is enclosed in the rhomboid

$$G := \{(x, y) \mid c \leq x \leq d, g_1(x) \leq y \leq g_2(x), g_3(x) \leq y \leq g_4(x)\}$$

where

g_1 is the line with slope q through $(a, p^{(i)}(a))$,
g_2 is the line with slope z through $(a, p^{(i)}(a))$,
g_3 is the line with slope q through $(b, p^{(i)}(b))$,
g_4 is the line with slope z through $(b, p^{(i)}(b))$.

The assumptions (a) and (b) are clearly designed for the case that Y is the non-monotonicity interval $[y_l, x_{l+1}]$ and that $0 \in p^{(i)}([y_l, x_{l+1}])$ thus indicating a suspicion of a zero in that interval.

The intersection of the lines g_1 and g_3 is denoted by $S_1 = (c_1, t_1)$ and by $S_2 = (c_2, t_2)$ the intersection of g_2 and g_4. If the interval $[t_1, t_2]$ does not contain a zero then there is no zero of $p^{(i)}$ in $[y_l, x_{l+1}]$. Otherwise the intersection of the rhomboid with the real axis is calculated. The bounds r and s for the slopes of $p^{(i-1)}$ in $[u_0, u_1]$ are obtained from the intersection of the intervals $p^{(i)}([u_0, u_1])$ and $[t_1, t_2]$.

Remark 6.6 A complete ALGOL program of the method is provided in

Dussel–Schmitt (1970). These authors report that although their method gives good results, it is fairly expensive. A comparison with other methods is not given.

6.5 THE METHOD OF HANSEN

The method described in Hansen (1979, 1981) combines several principles into a very efficient method for estimating the global minimum of functions of one or several variables. These functions are assumed to be twice continuously differentiable and to have only a finite number of stationary points. The method may also be applied if the above conditions are relaxed. In that case, however, it loses its effectiveness.

Hansen developed his ideas on the basis of the tools of optimization theory. This theory is concerned with stationary points of functions f and it applies gradient methods to obtain these points as well as the global minimum of f. One of Hansen's basic ideas is therefore to apply the interval Newton method to the derivative f' in subintervals $Y \subseteq X$, where X is the domain of f. In this manner a subset $S \subseteq Y$ is generated containing *all* the zeros of f'. The remainder of Y, $Y\setminus S$, can therefore be discarded since $Y\setminus S$ cannot contain the global minimum point (except for the case that the global minimum point occurs at the edge of X). This idea is now linked to the checking of the concavity of f over the subintervals Y using the test condition $F''(Y) < 0$. This is explained later. If indeed it is the case that $f'' < 0$ then it follows that the global minimum cannot occur at an interior point of Y. The checking of concavity does not result in an additional computational effort since outer estimates $F''(Y)$ of $f''(Y)$ are known from the application of the interval Newton method over Y.

In this section we describe Hansen's method in detail for the one-dimensional case following Hansen (1979). We then give a short outline of the multidimensional case. In this connection it should be noted that the main steps of the algorithm remain unchanged when going from the one- to the multidimensional case.

The basic premise for the method is that a global minimum is to be computed on a compact interval $X = [a, b] \in I$ for a function f that is twice continuously differentiable on X and where f' has at most finitely many zeros in X. This, together with the application of the algorithm to $-f$, constitutes a computation of an outer estimate for the range of f over X.

The algorithm is now described in detail.

A first estimate of the lower bound is obtained by computing

$$l = \min(f(a), f(b)).$$

If l^* is the global minimum in X then clearly $l \geq l^*$.

The algorithm proceeds by dynamically subdividing the interval X into subintervals. In order to proceed it is then assumed that including estimates F, F' and F'' can be computed for f, f' and f'' for each $Y \in I(X)$. These estimates cannot only be based on the present technique, but must also include one or more of the methods considered in the other sections of this monograph.

As the algorithm proceeds, f is evaluated at various points x of X. This is, for example, the case as a new subinterval Y is generated when an estimate $F(Y)$ is computed using the centred form (recall that in this case $f(c) = f(m(Y))$ must be computed). Furthermore, let l denote the currently smallest value of f found so far.

Three types of subintervals are now deleted from consideration.

(1) $Y \subseteq X$ is deleted if $0 \notin F'(Y)$
(2) $Y \subseteq X$ is deleted if $F(Y) > l$ since this implies that min $F(Y) > l$
(3) $Y \subseteq X$ is deleted if max $F''(Y) < 0$.

These intervals are deleted since the global minimum cannot occur at points in these intervals with the exception of the endpoints a or b in X. If $f(a)$ or $f(b)$ is the global minimum, however, then the interval Y under consideration can still be discarded since the information that the minimum is the one or the other of these values is retained in the current values of l. The calculation of $F(Y)$ is furthermore expedited using the estimation $F''(Y)$ in a second order Taylor-form (see Section 3.5).

If the interval $Y \subseteq X$ is not deleted then $\tilde{f}(Y)$ may potentially contain the global minimum and two courses of action are possible. Either the interval Y is further subdivided or the interval Newton method (Moore, 1966) is applied to f' over Y in order to estimate including intervals for f' in Y. This procedure is given by the computation of

$$Y_0 = Y,$$
$$N(Y_n) = m(Y_n) - f'(m(Y_n))/F''(Y_n) \quad \text{if } 0 \notin F''(Y_n),$$
$$Y_{n+1} = Y_n \cap N(Y_n), \quad n = 0, 1, 2,\ldots$$

It was observed in Alefeld (1968) and Hansen (1978a) that the problems arising from $0 \in F''(Y_n)$ may be circumvented by considering $N(Y_n)$ to be an unbounded interval that is reduced by the intersection operation to either an empty set, a single interval or two intervals.

This therefore provides for a procedure for the computation of a set S consisting of intervals containing all the zeros of f' in Y. The complement of S with respect to Y will contain no zeros of f' and hence no points where f can have a local or a global minimum (with the possible exception of the points a and b) and it may therefore be deleted. A convergence theorem for this Newton-type iteration is found in Hansen (1978b).

In summary: the whole process starts, as mentioned above, by setting Y

= X. Each of the above steps are in turn applied to Y, either eliminating Y or parts of Y. The subintervals of X that remain to be considered as candidates for containing the global minimum are entered on a list and the process is repeated with the subinterval of maximal width.

Initially, the number of subintervals on the list can tend to grow rapdily. At a certain point in the calculations the intervals will become small enough so that the overestimations in the calculations of F, F' and F'' are small and either one new interval is calculated or the interval is eliminated completely. At the end of the calculations the number of intervals will have been reduced to small intervals containing the points where f takes on the value of the global minimum.

The method now suggests a number of strategic decision criteria:

(a) It is best to choose the largest interval on the list as the next interval. This is because we wish to home in on l^* quickly and the smaller l is the larger portion of an interval Y that we will be able to delete using criterion (2) of the deletion list.

(b) If f, f' and f'' are expensive to evaluate then it pays to search for the next largest interval (see option (a)). If f, f' and f'' are not expensive to evaluate then the search itself may be too time-consuming.

(c) If f, f' and f'' are expensive to evaluate then all the previous techniques should be applied at each step.

Hansen discusses several termination criteria for his algorithm. They depend on the actual purpose of the optimization of Hansen (1979). We mention only one which terminates the algorithm if

$$l - l^* \leqslant \varepsilon$$

where l is again the current minimum of all the values which occur at the computation. Since the global minimum, l^* is not known the condition $l - l^* \leqslant \varepsilon$ is replaced by

$$l - p \leqslant \varepsilon$$

where p is the minimum of the left endpoints of intervals $F(Y)$ which are supposed to contain the global minimum. If all such intervals are discarded by the algorithm, then

$$l = \min \tilde{f}(X).$$

At a numerical realization of the algorithm, rounding errors make a modification of this criterion necessary. This can be found in Hansen (1979).

Example 6.6 In Shubert (1972) the global minimum is computed for the function

$$f(x) = -\sum_{k=1}^{5} k \sin\left[(k+1)x + k\right]$$

over $[-10, 10]$. This function has period 2π and global minimum at three points x_1, x_2 and x_3 (both these facts will not be used at the application of the algorithm). Using a *slight modification* of the above method, Hansen (1979) required about 77 evaluations of f, f' and F'' to obtain

$x_1 \in [-6.7745\ 76144, -6.7745\ 76143]$,

$x_2 \in [\ \ 5.7917\ 89015,\ \ \ 5.7917\ 89064]$,

$x_3 \in [-4.9139\ 21876, -4.9139\ 21811]$,

$f(x_1) \in [-12.03124944, -12.03124943]$,

$f(x_2) \in [-12.03124945, -12.03124943]$,

$f(x_3) \in [-12.3124944, -12.3124943]$.

The method by Shubert required 444 evaluations of f as well as a knowledge of a Lipschitz constant for f.

The extension of this method to m dimensions involves the same principles, but considerably more complicated machinery than that used for the one-dimensional case. The extended algorithm used the second order Taylor method as described in Section 3.5. The interval Newton method is similarly developed for m dimensions taking into consideration the cases where a singular matrix technically should have been inverted. A complete numerical example is also presented in Hansen (1980).

The complete description of the m-dimensional case of the algorithm is not included here, but it is found in Hansen (1980). A few supplements are given by Di-jan–Yuo-kang (1983).

6.6 CENTRED FORMS AND INTERVAL OPERATORS

In order to obtain a solution ξ of an equation $f(x) = 0$, where $f:R^m \to R^m$, one may use the so-called Newton-transform $Q(x) = x - af(x)$, where a is a non-singular matrix. In Krawczyk (1982) centred forms of the Newton-transform are investigated and it is noticed that some of these forms yield known interval operators for solving $f(x) = 0$. Among these are found the operators used by Moore (1966), Nickel (1971, 1981), Hansen (1968, 1978b), Hansen–Sengupta (1981), Krawczyk (1969), Krawczyk–Selsmark (1980), Alefeld–Herzberger (1970), Wolf (1980) and Qi (1981).

Although the theory developed in Krawczyk (1982) is of great interest we omit discussing it in detail here since it relates mainly to iteration methods rather than to the range of a function. The connection to the general theory of centred forms nevertheless justifies a short discussion of Krawczyk's ideas via an example in order to give the reader an impression of the type of results that are obtainable.

In this section it is assumed that the centred forms are defined for vector-valued functions ϕ, $f: D \to R^m$ where $D \subseteq R_m$. This is simply obtained by applying Definition 3.1 m times (see also Remark 3.2). The reason for this slight modification is that the usual iteration methods for systems of equations on R^m are defined as vector iterations rather than componentwise iterations.

Let first

$$f: D \subseteq R^m \to R^m$$

be a function on D and let

$$f(x) = 0 \tag{6.27}$$

be the corresponding system of equations. The following simple definitions are required to establish some elementary facts for operator iterations. An extensive treatment of this topic is found in Ortega–Rheinboldt (1970).

Definition 6.5 The vector ξ in R^m is a solution of (6.27) iff $f(\xi) = 0$.

Definition 6.6 The vector $\xi \in R^m$ is a fixed point of f iff $f(\xi) = \xi$.

Definition 6.7 The Newton-transform of f is defined by

$$\phi(x) = x - af(x) \tag{6.28}$$

where $a \in R^{m \times m}$ is a non-singular matrix.

Remark 6.7 It is obvious that ξ is a solution of (6.27) iff ξ is a fixed point of ϕ. This means that Definition 6.7 provides a method for generating Newton-like iteration methods, hence the name.

Remark 6.8 Other operators for solving $f(x) = 0$ may be generated not using the Newton-transform. In the one-dimensional case one has for example

$$\phi_1(x) = x - a[f(x)]^2,$$
$$\phi_2(x) = x - a f(x)x.$$

In order to obtain a centred form for ϕ we now proceed as in Section 4.1. Given $X \in I(D)$ a function

$$s: X \times D \to R^m$$

is defined by

$$\begin{aligned} s(x, c) &= x - af(x) - c + af(c) \\ &= (x - c) - a(f(x) - f(c)). \end{aligned}$$

The general centred form now requires us to find functions

$$S: I(X) \to I$$

as well as functions

$$G^\rho: I(X) \to I^{m \times m}, \qquad \rho = 1, 2, \ldots, r$$

such that

$$s(x, \alpha(Y)) \subseteq S(Y) \subseteq \sum_{\rho=1}^{r} G^\rho(Y)(Y - \alpha(Y)) \qquad \text{for } Y \in I(X), x \in Y$$

where $\alpha: I(X) \to X$ is the developing point function.

Suppose now that there exists a function $G^*: I(X) \to I^{m \times m}$ such that

$$f(x) - f(\alpha(Y)) \in G^*(Y)(Y - \alpha(Y)) \qquad \text{for all } x \in Y \text{ and } Y \in I(X).$$

Then, setting $r = 1$, we define

$$G(Y) = G^1(Y) = e - G^*(Y)$$

where e is the $m \times m$ unit-matrix. From this it follows that

$$\begin{aligned} s(x, \alpha(Y)) &= (x - \alpha(Y)) - a(f(x) - f(\alpha(Y))) \\ &\in (e - G^*(Y))(Y - \alpha(Y)) \\ &= G(Y)(Y - \alpha(Y)). \end{aligned}$$

The conditions of Definition 3.1 are thus satisfied. Letting therefore

$$S(Y) = G(Y)(Y - \alpha(Y))$$

we obtain

$$\begin{aligned} \Phi(Y) &= \phi(\alpha(Y)) + G(Y)(Y - \alpha(Y)) \\ &= \alpha(Y) - af(\alpha(Y)) + (e - G^*(Y))(Y - \alpha(Y)) \end{aligned} \qquad (6.29)$$

which is therefore a centred form function for ϕ. Depending on the choice of $G^*(Y)$ as well as on the non-singular transform matrix a we obtain different interval operators which are appropriate for the establishment of properly convergent interval iteration procedures.

An interval iteration procedure for X generally consists of an interval operator $T: X \to X$ and an initial interval Y_0 containing the fixed point ξ. A sequence $\{Y_i\}_{i=0}^{\infty}$ of intervals is generated via $Y_{i+1} = T(Y_i) \cap Y_i$, $i = 0, 1, \ldots$, where the intersection is formed in order that the sequence shall satisfy

$$Y_0 \supseteq Y_1 \supseteq Y_2 \supseteq \ldots.$$

An extensive discussion of such methods is found in Alefeld–Herzberger

(1983), see also Alefeld–Herzberger (1970), Krawczyk (1980a, 1980b), and Krawczyk–Selsmark (1980).

The speed of convergence of an iteration procedure is dependent on the improvement gained in each step. Having used the centred form in developing (6.29) we know that the convergence is favourable since the centred form is quadratically convergent in many cases.

The use of the Newton-transform also makes it possible to make unified existence statements for fixed points of interval operators (see Krawczyk, 1982).

We now outline some examples of interval iteration methods based on the above idea.

Example 6.7 Let $X \in I^m$, $f: X \to R^m$ and assume that $a(x) = [f'(x)]^{-1}$ exists where $f'(x)$ is the Jacobian matrix of f at x. Furthermore let $L(Y)$ be an inclusion of $\tilde{f}'(Y)$ for each $Y \in I(X)$. Then define

$$G^*(Y) = a(\alpha(Y))\, L(Y).$$

Inserting this into (6.29) we obtain

$$\Phi(Y) = \alpha(Y) - a(\alpha(Y))\, f(\alpha(Y))$$
$$+ (e - a(\alpha(Y)))\, L(Y)\, (y - \alpha(Y)) \qquad (6.30)$$

This is the centred form function of $\phi(x) = x - a(x)f(x)$ and it is known as the *Krawczyk operator*. This operator has the advantage that only point matrices need to be inverted. The corresponding interval iteration procedure may be simplified by keeping $a(x)$ constant (see Krawczyk, 1982).

Example 6.8 Let $X \in I^m$ and let $f: X \to R^m$ satisfy an interval Lipschitz condition of the form

$$f(x) - f(z) \in L(Y)(x - z) \text{ for all } Y \in I(X) \text{ and } x, z \in Y$$

where $L(Y) = [l_1(Y), l_2(Y)]$ is an $m \times m$ interval matrix (see Krawczyk, 1982). Assume that for each $Y \in I(X)$ a non-singular $m \times m$ matrix $a(Y)$ exists such that $a^{-1}(Y) \in L(Y)$. Assume further that $|a(Y)|$ is the matrix the components of which are the absolute values of the components of $a(Y)$ and define

$$a^+(Y) = [|a(Y)| + a(Y)]/2$$
$$a^-(Y) = [|a(Y)| - a(Y)]/2$$

then all $|a(Y)| = a^+(Y) + a^-(Y)$ and $a^+(Y)$ and $a^-(Y)$ are non-negative matrices. Naturally, L, l_1, l_2, a^+, a^-, and a are operators and we may therefore define new operators l, g, g_1, g_2, and G on $I(X)$ by

$l(Y) \in L(Y)$ arbitrary,

$g = e - al$, that means $g(Y) = e - a(Y)l(Y)$,

$g_1 = e - a^+ l_2 + a^- l_1$,

$g_2 = e - a^+ l_1 + a^- l_2$,

$G = [g_1, g_2]$

We want to show that $g_1(Y) \leq g(Y) \leq g_2(Y)$. The right-hand inequality follows since

$$g(Y) = e - a(Y)l(Y) = e - [a^+(Y) - a^-(Y)]l(Y)$$
$$\leq e - a^+(Y)l_1(Y) + a^-(Y)l_2(Y) = g_2(Y).$$

Similarly for the left-hand inequality.

The centred form function

$$\Phi(Y) = \alpha(Y) - a(\alpha(Y)) + G(Y)(Y - \alpha(Y))$$

is then the interval operator described in Krawczyk–Selsmark (1980). As in Example 6.7 the matrix $a(\alpha(Y))$ can be kept constant when defining the corresponding interval iteration procedure.

6.7 REMAINDER AND INTERPOLATION FORMS

A very interesting concept was introduced by Cornelius–Lohner (1983). If $X \in I, f: X \to R$, and $Y \in I(X)$ is given, and if $\tilde{f}(Y)$ is to be approximated by an outer approximation, then f is represented in the form

$$f = g + r \tag{6.31}$$

where $\bar{g}(Y)$ can be computed exactly — provided the rounding errors are neglected. If $S(Y) \in I$ is an inclusion for $\bar{r}(Y)$, then

$$F(Y) = \bar{g}(Y) + S(Y) \tag{6.32}$$

is an inclusion for $\tilde{f}(Y)$. If g is constructed using Hermite interpolation and $F: I(X) \to I$ is considered to be an inclusion function for $\tilde{f}$ over X then, theoretically, any order k of convergence of F to $\tilde{f}$ is possible. In practice, however, $k \leq 4$, can be obtained with moderate effort and $k = 5$ or $k = 6$ with more effort. An inclusion function $F: I(X) \to I$ for $\tilde{f}$ is called *convergent (to $\tilde{f}$) of order k* if

$$w[F(Y)] - w[\tilde{f}(Y)] = O(w(Y)^k) \quad \text{for all } Y \in I(X).$$

We now discuss these ideas following the paper of Cornelius–Lohner (1983). This paper presents several realizations of possible procedures, explicit expressions for third-order approximations, connections with the mean-value form, and finally, numerical examples.

Let $Y \in I(X)$ be fixed at first and let the continuous function $f: X \to R$ have a representation of the form (6.31) with *continuous* functions $g, r: X \to R$. Let $S(Y) \in I(X)$ and $S(Y) \supseteq \bar{r}(Y)$. The function g can be interpreted as an approximation of f where r is the corresponding remainder term. The interval $S(Y)$ is an estimation of this remainder term in the sense that

$$r(x) \in S(Y) \quad \text{for } x \in Y. \tag{6.33}$$

The interval $F(Y)$ defined by (6.32) is then called a *remainder form of f on Y*. The corresponding assignment $Y \to F(Y)$ gives rise to a function $F: I(X) \to I$ which is called a *remainder form function* (or abbreviated, a *remainder form*) of f on X. The use of $\bar{g}$, however, implies that we can choose only very simple functions in practical applications, i.e. polynomials of degree at most 5, or monotone functions.

Theorem 6.13 *If $F(Y)$ is a remainder form of f on Y, then it follows that*

(i) $\bar{f}(Y) \subseteq F(Y)$,

(ii) $|\bar{f}(Y), F(Y)| \leq w[S(Y)] \leq 2|S(Y)|$.

Proof (i) From (6.31) and (6.33) it follows that

$$f(x) = g(x) + r(x) \in \bar{g}(Y) + S(Y)$$

for any $x \in Y$.

(ii) Since f and g are continuous and since Y is compact, there exist points $x_*, x^* \in Y$ and $y_*, y^* \in Y$, where f or g takes its minimum and its maximum, that is

$$\bar{f}(Y) = [f(x_*), f(x^*)],$$
$$\bar{g}(Y) = [g(y_*), g(y^*)].$$

setting $S(Y) = [s, t]$ we get

$$|\bar{f}(Y), F(Y)| = |\bar{f}(Y), \bar{g}(Y) + S(Y)|$$
$$= \max\{|f(x_*) - g(y_*) - s|, |f(x^*) - g(y^*) - t|\}.$$

Estimating the arguments separately yields

$$|f(x_*) - g(y_*) - s| = f(x_*) - g(y_*) - s \leq f(y_*) - g(y_*) - s$$
$$\leq [g(y_*) + t] - g(y_*) - s = t - s = w[S(Y)],$$

and similarly,

$$|f(x^*) - g(y^*) - t| \leq w[S(Y)],$$

from which the final estimation, i.e.

$$|\tilde{f}(Y), F(Y)| \leq w[S(Y)] \leq 2|S(Y)|$$

is obtained. The last inequality is justified by (1.6). □

It is important to use simple functions g for numerical applications of Theorem 6.13 in order to be able to compute $\bar{g}(Y)$. Furthermore, the remainder r should be small enough such that small values of $w[S(Y)]$ are obtained and thus a good approximation $F(Y)$ to $\tilde{f}(Y)$. Interpolation and Taylor polynomials of low degree are well suited for this purpose. Both are special cases of the Hermite interpolation polynomials.

Let integers $s \geq 0$, $n \geq 0$, and $m_0,\ldots,m_n > 0$ be given such that

$$s + 1 = m_0 + \ldots + m_n. \tag{6.34}$$

Furthermore, let $x_0,\ldots,x_n \in Y$ be $n + 1$ distinct points and f be $s + 1$ times differentiable in Y. Then p_s shall denote the uniquely defined interpolation polynomial of degree s satisfying the Hermite interpolation conditions

$$p_s^{(j)}(x_i) = f^{(j)}(x_i) \quad \text{for } j = 0, 1,\ldots,m_i - 1,$$

$$i = 0, 1,\ldots,n.$$

It is known that for any $x \in Y$ there exists a $\xi = \xi(x) \in Y$ such that

$$f(x) = p_s(x) + \frac{1}{(s + 1)!} f^{(s+1)}(\xi) \prod_{i=0}^{n} (x - x_i)^{m_i}$$

is a representation of f of the form (6.31). If $F^{(s+1)}: I(X) \to I$ is a *bounded inclusion function* for $f^{(s+1)}$, then the *first interpolation form (function)* $V_s: I(X) \to I$ for f according to the above assumptions is defined by

$$V_s(Y) = \bar{p}_s(Y) + \frac{1}{(s + 1)!} F^{(s+1)}(Y) \prod_{i=0}^{n} (Y - x_i)^{m_i}.$$

If $F^{(s+1)}$ even satisfies a *Lipschitz condition* and if there exists a function $\eta: I(X) \to R$ with

$$\eta(Y) \in F^{(s+1)}(Y)$$

then a continuous function $q_{s+1}: X \to R$ is defined to each $Y \in I(X)$ by

$$q_{s+1}(x) = p_s(x) + \frac{\eta(Y)}{(s + 1)!} \prod_{i=0}^{n} (x - x_i)^{m_i}.$$

(The dependence on Y is given via $\eta(Y)$, the x_i and p_s. The indication of this dependence is, however, suppressed in the notation for the functions q_{s+1}.) Therefore, for each $Y \in I(X)$ there exists a representation of f of the form (6.31) given by

Sec. 6.7] **Remainder and Interpolation Forms** 151

$$f(x) = q_{s+1}(x) + \frac{1}{(s+1)!} [f^{(s+1)}(\xi) - \eta(Y)] \prod_{i=0}^{n} (x - x_i)^{m_i}.$$

Depending on the Y that is chosen the *second interpolation form* of f on Y is defined as the interval

$$Y_s(Y) = \tilde{q}_{s+1}(Y) + \frac{1}{(s+1)!} [F^{(s+1)}(Y) - \eta(Y)] \prod_{i=0}^{n} (Y - x_i)^{m_i}.$$

The function $U_s: I(X) \to I$ is then called the *second interpolation form function* (abbreviated, *second interpolation form*) of f on X.

The following theorem shows that V_s and U_s are inclusion functions for $\tilde{f}$ of order $s + 1$ or $s + 2$.

Theorem 6.14 *Let the above assumptions for the first and second interpolation forms hold. Then, for any $Y \in I(X)$, we get*

(i) $\tilde{f}(Y) \subseteq V_s(Y)$,

$\tilde{f}(Y) \subseteq U_s(Y)$.

Furthermore, there exist constants α and β such that for any $Y \in I(X)$ we get

(ii) $w[V_s(Y)] - w[\tilde{f}(Y)] \leq \alpha w(Y)^{s+1}$

$w[U_s(Y)] - w[\tilde{f}(Y)] \leq \beta w(Y)^{s+2}$.

Proof Since both $V_s(Y)$ and $U_s(Y)$ are special cases of the remainder form (6.32) it follows immediately from Theorem 6.13 that (i) must hold. Now to the proof of (ii). In the case of $V_s(Y)$ the term $S(Y)$ has the form

$$S(Y) = \frac{1}{(s+1)!} F^{(s+1)}(Y) \prod_{i=0}^{n} (Y - x_i)^{m_i}.$$

If K is an upper bound for the real number $|F^{(s+1)}(Y)|$ for $Y \in I(X)$ (which exists by assumption), then we set

$$\alpha = \frac{4K}{(s+1)!}$$

Using (1.10), Theorem 6.13(ii), (1.6), (1.6), (1.3), and (6.34) in the order given, we get the asserted estimation,

$$w[V_s(Y)] - w[\tilde{f}(Y)] \leq 2|V_s(Y), \tilde{f}(Y)| \leq 4|S(Y)|$$

$$= \frac{4}{(s+1)!} \left| F^{(s+1)}(Y) \prod_{i=0}^{n} (Y - x_i)^{m_i} \right|$$

$$= \frac{4}{(s+1)!} |F^{(s+1)}(Y)| \prod_{i=0}^{n} |Y - x_i|^{m_i}$$

$$\leq \alpha \prod_{i=0}^{n} w(Y - x_i)^{m_i} = \alpha \prod_{i=0}^{n} w(Y)^{m_i} = \alpha w(Y)^{s+1}.$$

In the case of $U_s(Y)$ the term $S(Y)$ has the form

$$S(Y) = \frac{1}{(s+1)!} [F^{(s+1)}(Y) - \eta(Y)] \prod_{i=0}^{n} (Y - x_i)^{m_i}.$$

If K is a Lipschitz constant for $F^{(s+1)}$ (which exists by assumption) then we set

$$\beta = \frac{4K}{(s+1)!}$$

We get the asserted estimation similarly to the case $V_s(Y)$, where only the estimation of $|F^{(s+1)}(Y)|$ is to be replaced by

$$|F^{(s+1)}(Y) - \eta(Y)| \leq w[F^{(s+1)}(Y) - \eta(Y)]$$
$$= w[F^{(s+1)}(Y)] \leq Kw(Y). \qquad \square$$

Now we know that V_s is an inclusion function for $\tilde{f}$ of order $s+1$ and that U_s is an inclusion function of order $s+2$. For getting higher than quadratic convergence it is sufficient if V_s is used with quadratic interpolation or with a quadratic Taylor polynomial ($s=2$) or to use U_s with linear interpolation or a linear Taylor polynomial ($s=1$). Obviously in these cases the computation of $\bar{g}(Y)$ is easy since g is a quadratic polynomial.

Remark 6.9 If $0 \in F^{(s+1)}(Y)$ then $\eta(Y) = 0$ should be chosen for constructing $U_s(Y)$. In this case q_{s+1} reduces to p_s such that only $\bar{p}_s(Y)$ has to be computed.

Remark 6.10 The results can be improved in the same manner as for the standard centred form, using the extended arithmetic when the power products $P(Y) = \prod_{i=0}^{n} (Y - x_i)^{m_i}$ which occur in the definition of V_s and U_s are evaluated. There are two reasonable ways of using the extended evaluation. For this purpose we introduce the functions

$$\Pi_i(x) = (x - x_i)^{m_i} \quad \text{for } i = 0,\ldots,n,$$

$$\Pi(x) = \prod_{i=0}^{n} (x - x_i)^{m_i}.$$

Then

$$\bar{P}(Y) = \prod_{i=0}^{n} \overline{\Pi_i(Y)}$$

or even

$$\hat{P}(Y) = \overline{\Pi(Y)}$$

can be used instead of $P(Y)$. Let $\tilde{V}_s(Y)$ or $\tilde{U}_s(Y)$ and $\hat{V}_s(Y)$ or $\hat{U}_s(Y)$ denote the remainder forms which arise from $V_s(Y)$ and $U_s(Y)$ by replacing the occurring term $P(Y)$ by $\tilde{P}(Y)$ or $\hat{P}(Y)$. By (2.15) it is then clear that for any $Y \in I(X)$ we get

$$\tilde{f}(Y) \subseteq \hat{V}_s(Y) \subseteq \tilde{V}_s(Y) \subseteq V_s(Y),$$
$$\tilde{f}(Y) \subseteq \hat{U}_s(Y) \subseteq \tilde{U}_s(Y) \subseteq U_s(Y).$$

Example 6.9 *Determination of $U_1(Y)$ using Taylor expansion at $c \in Y$ ($s = 1, n = 0, x_0 = c$. $m_0 = 2$). Let $Y \in I(X)$ and $\eta = \eta(Y) \in F^{(2)}(Y)$ be given. In the case $\eta \neq 0$ we write*

$$f(x) = f(c) + f'(c)(x - c) + f''(\xi)(x - c)^2/2$$
$$= q_2(x) + [f''(\xi) - \eta](x - c)^2/2$$

where

$$q_2(x) = f(c) + [x - c + f'(c)/\eta]^2 \eta/2 - f'(c)^2/(2\eta),$$

and obtain

$$U_1(Y) = \tilde{q}_2(Y) + [F^{(2)}(Y) - \eta](Y - c)^2/2,$$
$$\hat{U}_1(Y) = \tilde{U}_1(Y) = \tilde{q}_2(Y) + [F^{(2)}(y) - \eta] [0, |Y - c|^2/2].$$

The range $\tilde{q}_2(Y)$ can easily be determined. If $\eta = 0$, we set

$$q_2(x) = p_1(x) = f(c) + f'(c)(x - c),$$

and $\tilde{q}_2(Y)$ can be determined at once and inserted in the previous formulas for $U_1(Y)$ and $\hat{U}_1(Y) = \tilde{U}_1(Y)$.

Example 6.10 *Determination of $U_1(Y)$ using linear interpolation at $c, d \in Y$ where $Y = [c, d]$ ($s = n = 1, x_0 = c < x_1 = d, m_0 = m_1 = 1$). Let $Y \in I(X)$ and $\eta = \eta(Y) \in F^{(2)}(Y)$ be given and let $\Delta = [f(d) - f(c)]/(d - c)$. Then, in the case $\eta \neq 0$ we write*

$$f(x) = f(c) + \Delta(x - c) + f''(\xi)(x - d)(x - c)/2$$
$$= f(c) + \Delta(x - c) + (x - d)(x - c)\eta/2$$
$$+ [f''(\xi) - \eta](x - d)(x - c)/2$$
$$= q_2(x) + [f'(\xi) - \eta](x - d)(x - c)/2$$

where

$$q_2(x) = \frac{f(c) + f(d)}{2} + \frac{\eta}{2}\left[\left(x - \frac{c + d}{2} + \frac{\Delta}{\eta}\right)^2 - \left(\frac{\Delta}{\eta}\right)^2 - \frac{(d - c)^2}{4}\right]$$

is such that $\tilde{q}_2(Y)$ can be determined easily. We therefore obtain

$$U_1(Y) = \bar{U}_1(Y) = \bar{q}_2(Y) + [F^{(2)}(Y) - \eta](Y - d)(Y - c)/2,$$
$$\hat{U}_1(Y) = \bar{q}_2(Y) + [F^{(2)}(Y) - \eta][-w(Y)^2/8, 0].$$

In the case of $\eta = 0$ the given rearrangement of f can be suppressed using Remark 6.9 as in Example 6.9. This yields

$$q_2(x) = p_1(x) = f(c) + \Delta(x - c)$$

and $\bar{q}_2(Y)$ can be determined at once and inserted into the previous formulas for $U_1(Y) = \bar{U}_1(Y)$ and $\hat{U}_1(Y)$.

Finally, we compare the mean value form with the second interpolation form U_0, where both forms are quadratically convergent. Let $Y \in I(X)$, $c \in Y$, $\eta \in F^{(1)}(Y)$, and $F^{(1)}(Y) \subseteq \bar{f}(Y)$. Let

$$F(Y) = f(c) + F^{(1)}(Y)(Y - c)$$

be the mean value form as developed in Section 3.5 and let

$$U_0(Y) = \bar{q}_1(Y) + [F^{(1)}(Y) - \eta](Y - c)$$

where $q_1(x) = f(c) + \eta(x - c)$. ($U_0(Y)$ is obtained by interpolating at $c \in Y$ using the constant $p_0(x) = f(c)$.) Now, because of subdistributivity it follows that

$$F(Y) = f(c) + [F^{(1)}(Y) - \eta + \eta](Y - c)$$
$$\subseteq f(c)\eta(Y - c) + [F^{(1)}(Y) - \eta](Y - c) = U_0(Y).$$

This estimation shows that the mean value form is never worse than U_0. The equality $F(Y) = U_0(Y)$ can, however, be forced if η is chosen in a suitable way. For example, if $F^{(1)}(Y) = [k, l]$ and

$$\eta = \begin{cases} k & \text{if } k > 0, \\ 0 & \text{if } k \leq 0 \leq l, \\ l & \text{if } l < 0, \end{cases}$$

then $F(Y) = U_0(Y)$ as can be easily checked. This equality is also obtained if $\eta = (k + l)/2$. Nevertheless, when the quadratic convergence property is needed then the mean value form is to be preferred to the interpolation form U_0 because of its lower computational complexity.

Numerical results. We show some interesting results presented in Cornelius–Lohner (1983). They compared the intervals $\hat{U}_1(Y)$ of Examples 6.9 and 6.10 with the mean value form. The forms $\hat{U}_1(Y)$ of Example 6.9 will be denoted by $\hat{U}_1^T(Y)$ in the following tables and the notation for the forms $\hat{U}_1(Y)$ of Example 6.10 will be kept. The computations were done in the programming language PASCAL-SC which is an extension of PASCAL with a maximum accuracy arithmetic.

These authors preferred to compare the data using the Hausdorff metrics. This does not matter because of (1.10). The inclusions $F^{(2)}(Y)$ are those intervals which are computed in PASCAL-SC (using the extended power evaluation) when the interval Y is assigned to the argument of a function declaration for f''.

Table 6.1

$f(x) = \dfrac{x^2 - 5x + 9}{x - 5}$, $\quad Y_i = 2 + 10^{-i}[-1, 1]$ for $i = 0,\ldots,7$, $\quad \eta(Y_i) = 2$ for $i = 0,\ldots,7$

i	$w(Y_i)$	$\|\tilde{f}(Y_i), F(Y_i)\|$	$\|\tilde{f}(Y_i), \hat{U}_1(Y_i)\|$	$\|\tilde{f}(Y_i), \hat{U}_1^T(Y_i)\|$
0	$2 \cdot 10^0$	$4 \cdot 10^0$	$8 \cdot 10^{-1}$	$6 \cdot 10^{-1}$
1	$2 \cdot 10^{-1}$	$2 \cdot 10^{-2}$	$4 \cdot 10^{-4}$	$3 \cdot 10^{-4}$
2	$2 \cdot 10^{-2}$	$2 \cdot 10^{-4}$	$3 \cdot 10^{-7}$	$3 \cdot 10^{-7}$
3	$2 \cdot 10^{-3}$	$2 \cdot 10^{-6}$	$4 \cdot 10^{-10}$	$3 \cdot 10^{-10}$
4	$2 \cdot 10^{-4}$	$2 \cdot 10^{-8}$	$2 \cdot 10^{-11}$	$1 \cdot 10^{-11}$
5	$2 \cdot 10^{-5}$	$2 \cdot 10^{-10}$	$2 \cdot 10^{-11}$	$1 \cdot 10^{-11}$
6	$2 \cdot 10^{-6}$	$2 \cdot 10^{-12}$	$2 \cdot 10^{-11}$	$1 \cdot 10^{-11}$
7	$2 \cdot 10^{-7}$	$1 \cdot 10^{-12}$	$2 \cdot 10^{-11}$	$1 \cdot 10^{-11}$

Table 6.2

$f(x) = \dfrac{x + 2}{\sqrt[y]{|\cdot|}}$, $\quad Y_i = 2 + 10^{-i}[-1, 1]$ for $i = 0,\ldots,7$, $\quad \eta(Y_i) = 2$ for $i = 0,\ldots,7$

i	$w(Y_i)$	$\|\tilde{f}(Y_i), F(Y_i)\|$	$\|\tilde{f}(Y_i), \hat{U}_1(Y_i)\|$	$\|\tilde{f}(Y_i), \hat{U}_1^T(Y_i)\|$
0	$2 \cdot 10^0$	$5 \cdot 10^{-1}$	$1 \cdot 10^0$	$9 \cdot 10^{-1}$
1	$2 \cdot 10^{-1}$	$2 \cdot 10^{-3}$	$2 \cdot 10^{-4}$	$2 \cdot 10^{-4}$
2	$2 \cdot 10^{-2}$	$2 \cdot 10^{-5}$	$2 \cdot 10^{-7}$	$2 \cdot 10^{-7}$
3	$2 \cdot 10^{-3}$	$2 \cdot 10^{-7}$	$2 \cdot 10^{-10}$	$2 \cdot 10^{-10}$
4	$2 \cdot 10^{-4}$	$2 \cdot 10^{-9}$	$5 \cdot 10^{-11}$	$2 \cdot 10^{-11}$
5	$2 \cdot 10^{-5}$	$2 \cdot 10^{-11}$	$6 \cdot 10^{-11}$	$1 \cdot 10^{-11}$
6	$2 \cdot 10^{-6}$	$1 \cdot 10^{-11}$	$5 \cdot 10^{-11}$	$1 \cdot 10^{-11}$
7	$2 \cdot 10^{-7}$	$1 \cdot 10^{-11}$	$4 \cdot 10^{-11}$	$1 \cdot 10^{-11}$

Table 6.3

$$f(x) = \frac{\log x}{x}, Y_i = 3/2 + 10^{-i}[-1, 1] \text{ for } i = 0,\ldots,7, \qquad \eta(Y_i) = 3/2 \text{ for } i = 0,\ldots,7$$

i	$w(Y_i)$	$\|\tilde{f}(Y_i), F(Y_i)\|$	$\|\tilde{f}(Y_i), \hat{U}_1(Y_i)\|$	$\|\tilde{f}(Y_i), \hat{U}_1^T(Y_i)\|$
0	$2\cdot 10^0$	$7\cdot 10^0$	$2\cdot 10^1$	$2\cdot 10^1$
1	$2\cdot 10^{-1}$	$1\cdot 10^{-2}$	$9\cdot 10^{-4}$	$7\cdot 10^{-4}$
2	$2\cdot 10^{-2}$	$1\cdot 10^{-4}$	$9\cdot 10^{-7}$	$6\cdot 10^{-7}$
3	$2\cdot 10^{-3}$	$1\cdot 10^{-6}$	$9\cdot 10^{-10}$	$6\cdot 10^{-10}$
4	$2\cdot 10^{-4}$	$1\cdot 10^{-8}$	$1\cdot 10^{-12}$	$1\cdot 10^{-12}$
5	$2\cdot 10^{-5}$	$1\cdot 10^{-10}$	$1\cdot 10^{-12}$	$1\cdot 10^{-12}$
6	$2\cdot 10^{-6}$	$2\cdot 10^{-12}$	$1\cdot 10^{-12}$	$2\cdot 10^{-12}$
7	$2\cdot 10^{-7}$	$1\cdot 10^{-12}$	$1\cdot 10^{-12}$	$2\cdot 10^{-12}$

Table 6.4

$$f(x) = \exp(x - \sin x) - 1, \quad Y_i = -3/2 + 10^{-i}[-1, 1] \text{ for } i = 0,\ldots,7,$$
$$\eta(Y_i) = -3/2 \text{ for } i = 0,\ldots,7$$

i	$w(Y_i)$	$\|\tilde{f}(Y_i), F(Y_i)\|$	$\|\tilde{f}(Y_i), \hat{U}_1(Y_i)\|$	$\|\tilde{f}(Y_i), \hat{U}_1^T(Y_i)\|$
0	$2\cdot 10^0$	$3\cdot 10^0$	$2\cdot 10^0$	$3\cdot 10^0$
1	$2\cdot 10^{-1}$	$1\cdot 10^{-2}$	$1\cdot 10^{-3}$	$8\cdot 10^{-4}$
2	$2\cdot 10^{-2}$	$1\cdot 10^{-4}$	$6\cdot 10^{-7}$	$8\cdot 10^{-7}$
3	$2\cdot 10^{-3}$	$1\cdot 10^{-6}$	$6\cdot 10^{-10}$	$8\cdot 10^{-10}$
4	$2\cdot 10^{-4}$	$1\cdot 10^{-8}$	$1\cdot 10^{-12}$	$2\cdot 10^{-12}$
5	$2\cdot 10^{-5}$	$1\cdot 10^{-10}$	$1\cdot 10^{-12}$	$2\cdot 10^{-12}$
6	$2\cdot 10^{-6}$	$2\cdot 10^{-12}$	$1\cdot 10^{-12}$	$2\cdot 10^{-12}$
7	$2\cdot 10^{-7}$	$1\cdot 10^{-12}$	$1\cdot 10^{-12}$	$2\cdot 10^{-12}$

Table 6.5

$$f(x) = (16x^2 - 24x + 5)\exp(-x), \quad Y_i = 2.9 + 10^{-i}[-1, 1]$$
$$\text{for } i = 0,\ldots,7, \quad \eta(Y_i) = 2.9 \text{ for } i = 0,\ldots,7$$

i	$w(Y_i)$	$\|\tilde{f}(Y_i), F(Y_i)\|$	$\|\tilde{f}(Y_i), \hat{U}_1(Y_i)\|$	$\|\tilde{f}(Y_i), \hat{U}_1^T(Y_i)\|$
0	$2\cdot 10^0$	$3\cdot 10^1$	$1\cdot 10^1$	$1\cdot 10^1$
1	$2\cdot 10^{-1}$	$1\cdot 10^{-1}$	$6\cdot 20^{-3}$	$7\cdot 10^{-3}$
2	$2\cdot 10^{-2}$	$9\cdot 10^{-4}$	$6\cdot 10^{-6}$	$6\cdot 10^{-6}$
3	$2\cdot 10^{-3}$	$9\cdot 10^{-6}$	$6\cdot 10^{-9}$	$6\cdot 10^{-9}$
4	$2\cdot 10^{-4}$	$9\cdot 10^{-8}$	$1\cdot 10^{-11}$	$2\cdot 10^{-11}$
5	$2\cdot 10^{-5}$	$9\cdot 10^{-10}$	$1\cdot 10^{-11}$	$2\cdot 10^{-11}$
6	$2\cdot 10^{-6}$	$3\cdot 10^{-11}$	$1\cdot 10^{-11}$	$3\cdot 10^{-11}$
7	$2\cdot 10^{-7}$	$4\cdot 10^{-11}$	$1\cdot 10^{-11}$	$3\cdot 10^{-11}$

APPENDIX

Each two continuous norms on I^m are equivalent

We show that continuous norms on I^m are equivalent. Thus I^m acts just like finite-dimensional vector spaces. As far as we have investigated, it is not possible to reduce this assertion to the norm equivalence in R^m since it has not been possible to show that any norm of I^m may be expanded to a norm on R^{2m}. The reader may feel that there are no connections between centred forms and theoretic norm properties of I^m such as the above. The reason for including this appendix is that we have tried to keep the present monograph as simple as possible. Because of the equivalence of the norms we can use the maximum norm in nearly all analytic respects without losing the generality. The advantage of this norm is the close connection to the Hausdorff metric and to the width of an interval and, finally, the computational simplicity. For example, the Krawczyk–Nickel theorem (Theorem 4.2) can be presented in a far simpler form than in the original version.

Just as in analysis, two norms $\| \|_1$ and $\| \|_2$ on I^m are called equivalent if there are positive real numbers c and d such that

$$c\|X\|_1 \leq \|X\|_2 \leq d\|X\|_1 \qquad \text{for all } X \in I^m. \tag{A.1}$$

We keep in mind that the Hausdorff metric induces the maximum norm, see (1.9) and that the convergence defined by the Hausdorff metric is component-wise, see Section 2.5. Since the maximum norm on I^m is continuous, the unit ball

$$B_m = \{X \in I^m : \|X\| \leq 1\}$$

is compact, that is, any sequence in B_m has a convergent subsequence. Furthermore, in the proof we require the fact that the maximum norm can be written as

$$|X| = |m(X)| + w(X)/2 \qquad \text{for } X \in I$$

where $m(X)$ denotes the midpoint and $w(X)$ the width of X.

Each two continuous norms on I^m are equivalent

Theorem A.1 *All continuous norms are equivalent in I^m.*

Proof It is shown that any continuous norm ψ on I^m is equivalent to the maximum norm. For this purpose we need the following vectors e_i and E_i of I^m ($i = 1,\ldots,m$),

$$e_i = (0,\ldots,1,\ldots,0) \quad \text{and} \quad E_i = (0,\ldots,[-1, 1],\ldots,0),$$

where all coordinates with the exception of the *i*th are 0. Then any interval $X = (X_1,\ldots,X_m) \in I^m$ may be represented as

$$X = \sum_{i=1}^{m} [m(X_i)e_i + w(X_i)E_i/2].$$

The existence of the positive number d of (A.1) is shown first when ψ corresponds to $\|\ \|_2$ and $\|\ \|$ to $\|\ \|_1$ and where $\psi_i = \max\{\psi(e_i), \psi(E_i)\}$:

$$\psi(X) \leq \sum_{i=1}^{m} [|m(X_i)|\psi(e_i) + w(X_i)\psi(E_i)/2]$$

$$\leq \sum_{i=1}^{m} [|m(X_i)| + w(X_i)/2]\psi_i$$

$$= \sum_{i=1}^{m} |X_i|\,\psi_i \leq \|X\| \sum_{i=1}^{m} \psi_i.$$

By the previous calculation, one can choose $\sum_{i=1}^{m} \psi_i$ as the desired constant d.

The existence of the positive number c of (A.1) must now be shown, that is, that there exists a $c > 0$ such that for any $X \in I^m$ the inequality $c\|X\| \leq \psi(X)$ holds. This is proven via a contradiction where we assume that for every positive real c, an interval exists such that $c\|X\| > \psi(X)$. For every natural number n we can then find an interval X_n such that

$$\|X_n\|/n > \psi(X_n). \tag{A.2}$$

Furthermore, $\|X_n\| > 0$ for all n. We set $Y_n = X_n/\|X_n\|$ such that $\|Y_n\| = 1$, and such that (A.2) is equivalent to

$$\|Y_n\|/n > \psi(Y_n) \quad \text{for all natural numbers } n. \tag{A.3}$$

Since $(Y_n)_{n=0}^{\infty}$ is a sequence in the compact ball B_m, a subsequence exists, also denoted by $(Y_n)_{n=0}^{\infty}$, that converges to an interval Y (in the sense of the $\|\ \|$- norm). since the norm ψ is also continuous in the space $(I^m, \|\ \|)$, we have $\lim_{n \to \infty} \psi(Y_n) = \psi(Y)$. From (A.3) we conclude $\psi(Y) = 0$, that is, $Y = o$. This is a contradiction to the fact that $\|Y\| = 1$ which follows from the continuity of $\|\ \|$. $\square$

Bibliography

Alefeld, G. (1968). Intervallrechnung über den komplexen Zahlen und einige Anwendungen. Dissertation, Universität Karlsruhe.
—— (1981). Bounding the slope of polynomial operators and some applications. *Computing*, **26**, pp. 227–237.
Alefeld, G. and Herzberger, J. (1970). Über das Newton-Verfahren bei nichtlinearen Gleichungssystemen, *ZAMM*, 50, pp. 773–774.
—— —— (1974). *Einführung in die Intervallrechung*. Bibliographisches Institut, Mannheim.
—— —— (1983). *Introduction to Interval Computations*. Academic Press, New York.
Alefeld, G. and Rokne, J. (1981). On the evaluation of rational functions in interval arithmetic, *SIAM Journal on Numerical Analysis*, 18, pp. 862–870.
Apostolatos, N. and Kulisch, U. (1967). Grundlagen einer Maschinenintervallarithmetik. *Computing*, **2**, pp. 89–104.
Asaithambi, N. S., Zuhe, S. and Moore, R. E. (1982). On computing the range of values. *Computing*, **28**, pp. 225–237.
Bernstein, S. (1912). Demonstration du théorème de Weierstrass, fondée sur le calcul des probabilites. *Commun. Soc. Math. Kharkow*, 2, pp. 1–2.
Bonnesen, T. and Fenchel, W. (1934). *Theorie der konvexen Körper*. Springer-Verlag, Berlin.
Caprani, O., and Madsen, K. (1980). Mean value forms in interval analysis. *Computing*, **25**, pp. 147–154.
Cargo, G. T. and Shisha, O. (1966). The Bernstein form of a polynomial, *J. Res. Nat. Bur. Stand.*, **70B**, pp. 79–81.
Cheney, E. W. (1966). *Introduction to Approximation Theory*, McGraw-Hill, New York.
Chuba, W. and Miller, W. (1972). Quadratic convergence in interval arithmetic, Part I, *BIT*, **12**, pp. 284–290.

Bibliography

Collatz, L. (1966). *Functional Analysis and Numerical Mathematics*. Academic Press, New York.

Cornelius, H. and Lohner, R. (1983). Computing the range of values of real functions with accuracy higher than second order. Manuscript, Karlsruhe.

Crane, M. A. (1975). A bounding technique for polynomial functions. *SIAM Journal on Applied Mathematics*, 29, pp. 751–754.

Di-Jan, G. and Yuo-Kang, F. (1983). Some interval tests on unconstrained global optimization. *Freiburger Intervall-Berichte* Nr. 83/1, Freiburg.

Dixon, L. C. W., Gomulka, J. and Szegö, G. P. (1975). Towards a global optimisation technique. In Dixon–Szegö, (1975), pp. 29–54.

— and Szegö, G. P. (ed.) (1975). *Towards global optimisation*. Proceedings of a workshop in Cagliari, 1974. North-Holland, Amsterdam.

— — (1978a). The global optimisation problem: an introduction. In Dixon–Szegö, (1978b), pp. 1–15.

— — (ed.) (1978b). *Towards global optimisation 2*, North-Holland, Amsterdam.

Dussel, R. and Schmitt, B. (1970). Die Berechnung von Schranken für den Wertebereich eines Polynoms in einem Intervall. *Computing*, 6, pp. 35–60.

Eggleston, H. G. (1966). *Convexity*. Cambridge University Press, Cambridge.

Gargantini, I. and Henrici, P. (1972). Circular arithmetic and the determination of polynomial zeros. *Numer. Math.*, 18, pp. 305–320.

Goldstein, A. J. and Richman, P. L. (1973). A midpoint phenomenon. *J. ACM*, 20, pp. 301–304.

— — (1975). A midpoint phenomenon, *SIAM Journal on Applied Mathematics*, 29, pp. 751–754.

Grant, J. A. and Hitchins, G. D. (1973). The solution of polynomial equations in interval arithmetic. *Computer Journal*, 16, pp. 69–72.

Grätzer, G. (1979). *Universal Algebra*. 2nd edition. Van Nostrand, Princeton, N.J.

Hansen, E. R. (1968). On solving systems of equations using interval arithmetic. *Mathematics of Computation*, 22, pp. 374–384.

— (ed.) (1969a). *Topics in Interval Analysis*. Oxford University Press.

— (1969b). The centered form. In *Topics in Interval Analysis*, ed. E. Hansen, Oxford, pp. 102–105.

— (1978a). A globally convergent interval method for computing and bounding real roots. *BIT*, 18, pp. 415–424.

— (1978b) Interval forms of Newton's method. *Computing*, 20, pp. 153–163.

— (1979). Global optimisation using interval analysis: the one-dimensional case. *Journal of Optimization Theory and Applications*, **29**, pp. 331–344.
— (1980). Global optimisation using interval analysis — the multi-dimensional case. *Numerische Mathematik*, **34**, pp. 247–270.
Hansen, E. R. and Sengupta, S. (1981). Bounding solutions of systems of equations using interval analysis. *BIT*, **21**, pp. 203–211.
Herzberger, J. (1977). Zur Approximation des Wertebereich reeller Funktionen durch Intervallausdrücke. In *Grundlagen der Computer Arithmetik*, ed. G. Alefeld, Springer-Verlag, pp. 57–64.
— (1978). A note on a bounding technique for polynomial functions. *SIAM Journal on Applied Mathematics*, **34**, pp. 685–686.
Hitchins, G. D (1972). An interval arithmetic package and some applications. *Technical Report* 10, Centre for Computer Studies, University of Leeds, England.
Hu, Sze-Tsen (1966). *Introduction to General Topology*, Holden Day Inc., San Francisco.
Krawczyk, R. (1969). Newton-Algorithmen zur Bestimmung von Nullstellen mit Fehlerschranken. *Computing*, **4**, pp. 187–201.
— (1980a). Interval extensions and interval iterations. *Computing*, **24**, pp. 119–129.
— (1980b). Zur Konvergenz iterierter Mengen. *Freiburger Intervall-Berichte*, 80/3, Freiburg.
— (1982). Zentrische Formen und Intervalloperatoren. *Freiburger Intervall-Berichte*, 82/1, Freiburg.
— (1983). Intervallsteigungen für rationale Funktionen und zugeordnete zentrische Formen. *Freiburger Intervall-Berichte*, 83/2, Freiburg.
Krawczyk, R. and Nickel, K. (1982). Die zentrische Form in der Intervallarithmetik, ihre quadratische Konvergenz und ihre Inklusionsisotonie. *Computing*, **28**, pp. 117–132.
Krawczyk, R. and Selsmark, F. (1980). Order convergence and iterative interval methods. *Journal of Mathematical Analysis*, **73**, pp. 180–204.
Kulisch, U. and Miranker, W. L. (1981). *Computer Arithmetic in Theory and Practice*, Academic Press, New York.
Lootsma, F. A. (ed.) (1972). *Numerical methods for non-linear optimization*. Conference in Dundee 1971. Academic Press, New York.
Lorenz, C. G. (1953). *Bernstein Polynomials*, University of Toronto Press, Toronto.
Marden, M. (1966). Geometry of Polynomials. *American Mathematical Society*. Rhode Island.
Markov, S. M. (1977). A differential calculus for interval-valued functions based on extended interval arithmetic. *C. R. Acad. Bulgare Sci*, **30**, pp. 1377–1380.

McCormick, G. P. (1972). Attempts to calculate global solutions of problems that may have local minima. *Numerical Methods for Non-linear Optimization*, ed. F. A. Lootsma, Academic Press, pp. 209–221.

Mendelson, E. (1964). *Introduction to Mathematical Logic*. Van Nostrand, Princeton, N.J.

Mihelcic, M. (1975). Eine Modifikation des Halbierungsverfahren zur Bestimmung aller reellen Nullstellen einer Funktion mit Hilfe der Intervall-Arithmetik, *Angew. Inform*, pp. 25–29.

Miller, W. (1972). Quadratic convergence in interval arithmetic, Part II. *BIT*, **12**, pp. 291–298.

— (1973a). More on quadratic convergence in interval arithmetic. *BIT*, **13**, pp. 76–83.

— (1973b). The error in interval arithmetic. *IBM Research*, RC 4338.

— (1975). The error in interval arithmetic. *Proceedings of The International Symposium*, ed. K. Nickel, Springer-Verlag, pp. 246–250.

Miranda, C. (1941). Un'osservasione su un teorema di Brouver. *Bol. Un. Mat. Ital.*, Serie II, **3**, pp. 5–7.

Moore, R. E. (1962). Interval Arithmetic and Automatic Error Analysis in Digital Computing, Ph.D. Thesis, Stanford University.

— (1966). *Interval Analysis*. Prentice-Hall, Englewood Cliffs, N.J.

— (1976). On computing the range of a rational function of n variables over a bounded region. *Computing*, **16**, pp. 1–15.

— (1979). *Methods and applications of interval analysis*, SIAM, Philadelphia.

Nickel, K. (1975) (ed.). *Interval Mathematics*, Proceedings of the International Symposium, Karlsruhe, 1975, Springer-Verlag, Berlin.

— (1976). Die vollautomatische Berechnung einer einfachen Nullstelle von $F(t) = 0$ einschliesslich einer Fehlerabschätzung, *Computing*, **2**, pp. 233–245.

— (1971). On the Newton method in interval analysis. *MRC Tech. Summary Report*, 1136, University of Wisconsin, Madison.

— (1977). Die Überschätzung des Wertebereiches einer Funktion in der Intervallrechnung mit Anwendungen auf lineare Gleichungssysteme. *Computing*, **18**, pp. 15–36.

— (1980) (ed.). *Interval Mathematics 1980*, Proceedings of the International Symposium on Interval Mathematics, Freiburg, 1980. Academic Press, New York.

— (1981). A globally convergent ball Newton method. *SIAM Journal on Numerical Analysis*, **18**, pp. 989–1003.

Ortega, J. M. and Rheinboldt, W. C. (1970). *Iterative Solution of Nonlinear Equations in Several Variables*. Academic Press, New York.

Petkovic, L. D. (1983). On two applications of Taylor series in circular complex arithmetic. *Freiburger Intervallberichte*, 83/2, Freiburg.

Qi, L. (1981). Interval boxes of solutions of nonlinear systems. *Computing*, **27**, pp. 137–144.

Raith, M. (1980). Existence and uniqueness of inclusion isotonic centred ball extensions. *Computing*, **24**, pp. 195–205.

Raith, M. and Rokne, J. (1982). Inclusion-isotone centred ball extensions. Manuscript.

Rall, L. B. (1981). *Automatic Differentiation: Techniques and Applications*. Springer, Berlin.

— (1983). Mean value and Taylor forms in interval analysis. *SIAM Journal on Math An.*, **14**, pp. 223–238.

Ratschek, H. (1971). Die Subdistributivität in der Intervallarithmetik. *Z. Angew. Math. Mech.*, **51**, pp. 189–192.

— (1974). Mittelwertsatz für Intervallfunktionen, *ZAMM*, **54**, T229–230.

— (1975). Nichtumerische Aspekte der Intervallarithmetik, *Interval Mathematics*, ed. K. Nickel, Springer-Verlag, pp. 48–74.

— (1977). Mittelwertsätze für Intervallfunktionen, *Beiträge zur Numerischen Mathmatik*, **6**, pp. 133–144.

— (1978). Zentrische Formen. *ZAMM*, **58**, T434–436.

— (1980a). Centred forms. *SIAM Journal on Numerical Analysis*, **17**, pp. 656–662.

— (1980b). Optimal approximations in interval analysis in *Interval Mathematics, 1980*, ed. K. Nickel, Academic Press, pp. 181–202.

Ratschek, H. and Rokne, J. (1980a). About the centred form. *SIAM Journal on Numerical Analysis*, **17**, 3, pp. 333–337.

—— (1980b). Optimality of the centered form. *Interval Mathematics 1980*, ed. K. Nickel, Academic Press, pp. 499–508.

—— (1981). Optimality of the centered form for polynomials. *Journal of Approximation Theory*, **32**, pp. 151–159.

Ratschek, H. and Schröder, G. (1971). Über die Ableitung von intervallwertigen Funktionen. *Computing*, **7**, pp. 172–187.

—— (1981). Centered forms for functions in several variables. *Journal of Mathematical Analysis and Applications*, **82**, pp. 543–552.

Rausch, T. (1981). Zentrische Formen. Fortgeschrittenen-Praktikum für Mathematiker. University of München.

Richman, P. L. (1969). Error control and the midpoint phenomenon. *Bell Telephone Laboratories* MM-69-1374-29.

Ris, F. N. (1972). Interval Analysis and Applications to Linear Algebra. D.Ph. Thesis, Oxford University, Oxford.

Rivlin, T. J. (1970). Bounds on a polynomial. *J. Res. Nat. Bur. Stand*, **74B**, pp. 47–54.

Rokne, J. (1977). Bounds for an interval polynomial. *Computing*, **18**, pp. 225–240.
— (1978). Polynomial least square interval approximation. *Computing*, **20**, pp. 165–176.
— (1979a). A note on the Bernstein algorithm for bounds for interval polynomials. *Computing*, **21**, pp. 159–170.
— (1979b). The range of values of a complex polynomial over a complex interval. *Computing*, **22**, pp. 153–169.
— (1981). The centered form for interval polynomials. *Computing*, **27**, pp. 339–348.
— (1982). Optimal computation of the Bernstein algorithm for the bound of an interval polynomial. *Computing*, **28**, pp. 239–246.
Rokne, J. and Grassmann, E. (1979). The range of values of a complex polynomial over a circular complex interval. *Computing*, **23**, pp. 139–169.
Rokne, J. and Wu, T. (1982a). The circular complex centered form. *Computing*, **28**, pp. 17–30.
— — (1983). A note on the circular complex centered form. *Computing*, **30**, pp. 201–211.
Shubert, R. O. (1972). A sequential method seeking the global maximum of a function, *SIAM J. on Numerical Analysis*, **9**, pp. 379–388.
Skelboe, S. (1974). Computation of rational interval functions. *BIT*, **14**, pp. 87–95.
— (1979). True worst-case analysis of linear electrical circuits by interval arithmetic. *IEEE Transactions on circuits and systems*, vol. CAS-26, pp. 874–879.
Spang, H. A. III (1962). A review of minimization techniques for nonlinear functions. *SIAM Rev.*, **4**, pp. 343–365.
Spaniol, O. (1970). Die Distributivität in der Intervallarithmetik. Computing, **5**, pp. 6–16.
Sunaga, T. (1958). Theory of an interval algebra and its application to numerical analysis. RAAG Memoirs, **2**, pp. 29–46.
Traub, J. F. and Woźniakowski, H. (1980). *A General Theory of Optimal Algorithms*. Academic Press, New York.
Wilansky, A. (1970). *Topology for Analysis*, Ginn, Waltham.
Wilde, D. J. (1978). *Globally Optimal Design*, Wiley, New York.
Wolf, M. A. (1980). A modification of Krawczyk's algorithm, *SIAM J. on Numer. Anal.*, **17**, pp. 376–379.

List of Symbols

I	13
R	14
I^m	15
o	15
$I(D)$	16
$A \cdot B$	16
$m(A)$	16
$w(A)$	16
R^+	17
$\|A\|$	17
$|A|$	17
$\vee$	18
$|A, B|$	18
$\bar{\bar{f}}$	21
$\hat{F}(X)$	39
$F_k, F_k(X)$	43
N	49

Index of Names

Alefeld, G.   13, 16, 17, 19, 31, 61, 66, 76, 111, 112, 142, 144, 146, 147, 159
Apostolatos, N.   24, 47, 159
Asaithambi, N. S.   93, 99, 102, 159

Bernstein, S.   133, 159
Bonnesen, T.   122, 159

Caprani, O.   76, 81, 159
Cargo, G. T.   130, 131, 132, 135, 136, 159
Cheney, E. W.   134, 159
Chuba, W.   31, 59, 66, 159
Collatz, L.   122, 160
Cornelius, H.   11, 148, 154, 160
Crane, M. A.   160

Di-Jan, G.   144, 160
Dixon, L. C. W.   9, 160
Dussel, R.   137, 141, 160

Eggelston, H. G.   160

Fenchel, W.   122, 159

Gargantini, I.   117, 160
Goldstein, A. J.   30, 160
Gomulka, J.   9, 160
Grant, J. A.   160
Grätzer, G.   23, 160
Grassmann, E.   164

Hansen, E. R.   30, 31, 35, 66, 99, 141, 142, 143, 144, 160, 161
Henrici, P.   117, 160
Herzberger, J.   13, 16, 17, 19, 31, 66, 76, 144, 146, 147, 159, 161
Hitchins, G. D.   160, 161
Hu, Sze-Tsen.   161

Krawczyk, R.   10, 19, 30, 31, 32, 37, 59, 61, 62, 63, 64, 66, 67, 69, 71, 72, 75, 76, 84, 88, 116, 129, 130, 144, 145, 147, 148, 157, 161
Kulisch, U.   13, 24, 47, 159, 161

Lohner, R.   11, 148, 154, 160
Lootsma, F. A.   9,. 161
Lorenz, C. G.   130, 161

Madsen, K.   76, 81, 159
Marden, M.   107, 161
Markov, S. M.   25, 161
McCormick, G. P.   9, 162
Mendelson, E.   23, 162
Mihelcic, M.   162
Miller, W.   31, 59, 66, 159, 162
Miranda, C.   66, 162
Miranker, W. L.   13, 161
Moore, R. E.   8, 12, 13, 14, 16, 17, 23, 24, 25, 26, 29, 30, 31, 35, 44, 55, 56, 57, 60, 66, 68, 75, 76, 78, 81, 91, 93, 94, 96, 98, 99, 102, 142, 144, 159, 162

Nickel, K.   13, 16, 19, 31, 63, 64, 66, 67, 69, 71, 72, 76, 92, 138, 144, 157, 161, 162

Ortega, J. M.   145, 162

Petkovic, L. D.   116, 163

Qi, L.   144, 163

Raith, M.   72, 76, 163
Rall, L. B.   56, 76, 163
Ratschek, H.   15, 19, 25, 30, 31, 40, 51, 53, 54, 62, 103, 110, 163
Rausch, T.   58, 163
Rheinboldt, W. C.   145, 162
Richman, P. L.   30, 160, 163

Index of Names

Ris, F. N. 15, 163
Rivlin, T. J. 130, 133, 135, 136, 163
Rokne, J. 31, 72, 103, 110, 111, 112, 116, 130, 135, 159, 163, 164

Schmitt, B. 137, 141, 160
Schröder, G. 25, 53, 54, 163
Selsmark, F. 144, 147, 148
Sengupta, S. 144, 161
Shisha, O. 130, 131, 132, 135, 136, 159
Shubert, R. O. 143, 144, 164
Skelboe, S. 57, 76, 90, 91, 93, 97, 99, 101, 102, 161, 164
Spang, H. A. 9, 164
Spaniol, O. 15, 164

Sunaga, T. 18, 164
Szegö, G. P. 9, 160

Traub, J. F. 103, 164

Wilansky, A. 14, 26, 164
Wilde, D. J. 9, 164
Wolf, M. A. 144, 164
Woźniakowski, H. 103, 164
Wu, T. 31, 116, 164

Yuo-Kang, F. 144, 160

Zuhe, S. 93, 99, 102, 159

Index of Subjects

absolute value 17
approximation 103
approximation, better 106
approximation, optimal 106

Bernstein approximation 133
Bernstein coefficients 131
Bernstein form 131
Bernstein functions 130
Bernstein polynomial 133

centred form 30, 65, 66, 119
centred form function 65
circular interval 116
computable from values 104
continuous 25
convergence 24
convergence of order k 148
cyclic bisection algorithm 102

dependence on values 104
developing point 65
developing point function 65
diameter included 121

expression 22
expression, arithmetic 22
expression, defining 22
expression, underlying 22
extended power evaluation 38

function corresponding to u_i 59
function procedure 59
function of a procedure 59

Hausdorff metric 18
Horner scheme, small 39
Horner scheme, evaluation 117

including approximation 103
inclusion function 65, 94
inclusion isotonicity 17, 24
interpolation form 150, 151
interpolation form function 150, 151
interval expression 22
interval function procedure 60
interval operations 13, 14
interval slope 60
interval vector 15

Index of Subjects

Krawczyk-operator 147
Krawczyk's centred form 59, 61, 75
Krawczyk's circular centred form 130

linearly convergent 66
Lipschitz condition 25
Lipschitz constant 25

maximum norm 17
mean-value form 78, 81
metric 18
metric, chain inclusion isotone 18
metric, homogeneous 18
metric, translation invariant 18
modulus 17
midpoint 16
multi-index 49

natural interval extension 23
nested form 39
Newton transform 145
norm 17
norm, inclusion isotone 17

occurs only once 91

point intervals 14
power sum evaluation 117

quadratic convergence 32, 66

range 21
range function 24
refinement 94
refinement, uniform 94
remainder form 149
remainder form function 149

simple power evaluation 38
Skelboe's method 99
standard centred form 50
standard centred form of order k 33, 43
standard circular centred form function 119
subdistributive law 15, 117
subdivision method 94
symmetric intervals 37

Taylor-form 77, 78, 79
Taylor-form function 77, 79

width 16